The
Great Pyramid
and
Connections
with
the Bible

PYRAMIDOLOGY V

ELO288

authorHOUSE®

AuthorHouse™
1663 Liberty Drive
Bloomington, IN 47403
www.authorhouse.com
Phone: 833-262-8899

Published by AuthorHouse 10/20/2021

ISBN: 978-1-6655-2370-7 (sc)
ISBN: 978-1-6655-2374-5 (e)

Library of Congress Control Number: 2021908263

Print information available on the last page.

This book is printed on acid-free paper.

Contents

Abstract

This essay is based on Adam Rutherford's [AR] work. His understanding was that the Great Pyramid [GP] represent "Bible in stone".

Measurement in the GP is 1 Inch = 1 year.

This has been 27 years, and three times calculation's into the future has been correct.

This essay has been done with Mathematics and Statistics, but also History and Theology. Statistics have confirmed the result within Science.

The main conclusion is that Pyramid specialists are mainly correct. Calculations for the time period 2000-2018 are correct, and now Science have the opportunity to do research concerning the time frame 2019-2060.

Útdráttur

Umfjöllunarefnið er í stuttu máli rannsóknarvinna Dr Adams Rutherford. Niðurstaða hans var sú, að Pýramídinn mikli væri "Biblía í steini".

Mælieining pýramídans mikla er 1 þumlungur = 1ár.

Eftir 27 ára rannsóknarvinnu hef ég þrívegis komið með tilgátur inn í framtíðina, sem reynst hafa verið rétt reiknaðar.

Mestmegnis hefur verið unnið með Stærðfræði og Tölfræði og tímabilið 2002-2018, að tilgátur hafa reynst vera réttar.

Nú hefur vísindasamfélagið tækifæri til að skoða tímabilið árin 2019-2060. Fjölmargar nýjungar koma fram í þessari ritgerð og Tölfræðin staðfestir þessar niðurstöður innan krafna vísinda.

Dedication

This PhD essay is dedicated to Pyramidology specialists.

Abbreviations

AR : Dr Adam Rutherford
GP: Great Pyramid
JT: John Taylor
PS: Dr Piazzi Smyth
RM: Robert Menzies
FB: Fred Binns
DD: David Davidson
HA: Dr Herbert Aldersmith
HKO: Haraldur Kristján Ólason
ET: Erlingur Þorsteinsson
SB: Sigurður Bjarnason

Acknowledgments

I am grateful for the assistant of Fred Binns, Sigurður Oddgeirsson, Guðni Guðnason, Þorsteinn G. Þorsteinsson, Sr. Ólafur Jóhansson at Grensáskirkja and to my family for understanding, support and patience.

A special thanks to my fellow Pyramid colleagues, Sigurður Bjarnason and Haraldur Kristján Ólason.

1

INTRODUCTION

Relatively few know that the Bible has a mathematical angle. Inside the Bible is clearly shown that behind is mathematical confirmation in numbers, also is mentioned in the Bible "That the matter would be sealed until the time of the end"

This was well known by the Jews who had their Cabala books that every letter has a number.

However, the true interpretation was to be reserved or "closed up and sealed till the time of the end". Dan 12: 9

Though this is a "final proof" of the Bible from Mathematical point of view, many have been doing this work within demands of science. The only missing puzzle is the End Time calculation. That puzzle is not the least important, and also the whole picture. I would say this part in the matter is around 5%, most results of these specialists have been correct, but because of its complexity like in scientific work, the hypothesis is set up and either they stand or fall.

The aim here is to concentrate on the work of Rutherford family, AR and James Rutherford.

Pyramidology's amongst other students of the Bible have been well aware of this method of interpretation, but have also noted the remarkable correlation between the Great Pyramid and the Bible. Using the numeric value of the passage in the Book of Isaiah which refers to this amazing structure they realized that this value is exactly equal to the height in inches of this very edifice. This passage reads "In that day shall there be an altar to the LORD in the midst of the land of Egypt and a pillar at the border thereof to the LORD. And it shall be for a sign and for a witness unto the LORD because of the oppressors, and he shall send a saviour, and a great one and shall deliver them" (Isaiah 19:19 -20)

This is just a single instance and simply included here as a demonstration of the value of this method of interpretation. This remarkable identification of the GP with the Biblical record was fully corroborated when it was also realised the length of the various passage systems corresponded, in inches to years, from the date of the pyramids construction to the exact year of the commencement of, what is now termed the Jewish dispensation, to its termination with the advent of Jesus Christ, (the saviour of the above text), the length of His ministry, and other historic events consequential to the advance of the Gospel Message. Very many other features have been discovered by various researchers since, but one compelling Question remains. Is the date of the End Time mentioned in the Scriptures concealed in this remarkable structure? The aim then of this present work is to further examine this remarkable structure with the object of determining any End Time features that may be present. Of course, even as with any scientific research, these projections will be to some degree hypothetical until validated by events. With regard to our method we will be basing our calculations, in the main on the work of Adam Rutherford.

Because of demands of scientific claims this in fact good news. Confirming that the timeframe 2002-2018, as many things have come through.

It is very gratifying to have this method of research which can stand up to scientific scrutiny as recent new investigation particularly in 2018 have located further End Time features.

Many things have been discovered in the year 2018. In Ante-Chamber the years 1844-2051 can be seen and 93 years of delay. We can also see the scored lines in the Great Pyramid 2141 B.C. which is the calculation point of the Great Pyramid, and 1440 A.C. is also shown in the Great Pyramid the year printing started, and therefore increasing of knowledge.

In November 2017, FB agreed to the new findings of the End Time calculation's, but added that Newton, had calculated this before with a different method. Some prophecies have not been fulfilled.

Why mentioning this? Concerning the Rosetta stone which was a puzzle understanding the text on the stone, for a long time there came up a quarrel between persons from France and Britain. Thomas Young in the year 1814, which was first to solve the matter. Correct is that the British person was first to solve the puzzle partly, but the French Jean Francois Champollion in the year (1822-1824) person solved the puzzle mostly, later the stone was solved fully and detailed information's concerning this matter is in history books. The Rosetta stone is from 196 B.C. and was found in the year 1799 is key to the past. This is also of importance concerning these matters, because the stone explains in three different languages and the Egypt world was opened for further understanding. Newton discovered from GP his three laws of motions from numbers from the GP. Newton likely thought that GP is divinely inspired. Talked about that he was "standing on the shoulders of Giants"

A hypothesis in the year 1993-4 concerning End Time 2002-2016/7, and 10 years shortage or 2006/7 and 2016,257 was found in the Ante-Chamber that year (2016). In both cases, something happened. The banks fell in 2008 which is admitted that it actually happened in the year 2006. And then 2016 Panama papers up to date 3-4 April 2016.

Concerning Science usual method is that experiments should be possible to repeat. Time is not possible to repeat to my knowledge, so time frame from 2019-2060 that is within the time frame of Newton certain things must happen within this time frame so science can accept these matters. This is quite important so disagreements should be able to settle.

Then FB took over working fulltime job in these matters. According to the GP, all that accept Jesus Christ are saved.

Just through this understanding, many religions are claimed as false by GP understanding. This is the result of the Pyramidology.

Many new findings are as breakthrough concerning Bible understanding, and according to FB is globally misunderstood.

HA's suggestions were 1928 to look at especially, but when these 93 years delay is shown in Ante-Chamber then the year 2021,135 has 93 years earlier specific meaning compared to 1928. In King Chamber 1928-1953 and the latter year 1953 was specially looked at by Pyramidology. The time difference is 25 years as between 2021- 2046 or also 25 years. This is a very recent finding and FB had never heard of such finding.

It is also the years 2021 to 2029, comparing to King Chamber years 1928-1936. This was especially under deep research of AR.

One of the scientific methods is to be able to do the test again, in these matters for the obvious reason that is not possible. Only God seems to be able to see through time.

Here in this booklet just mentioned names of most of them who took part in this complexity work. The main source derives from AR who published the book *Israel Britain*. Icelandic translation was made, it is called *Visindaleg opinberun*. The aim here is to correct AR findings, but also to confirm his work. AR got his Doctor degree for this work.

2

Origin of the Land Guardians

Iceland gained sovereignty in 1918, it was decided that a special national Icelandic coat of arms should be taken up. With a royal decree on February 12, 1919, the Coat of Arms was decided and should be "crowned shield and on the flag of Iceland. The land guardians are the well-known landlords like " Dragon, vulture, ox and giant, "says the ruling. Now, with the complete separation of Iceland from Denmark, the Coat of Arms has been decided again and completely the same as in 1918, unless the crown has disappeared as soon as the king's power. It is now rather strange that this should be the case, that these four old land guardians should enter the Coat of Arms of Iceland. We do not know much about the beliefs that followed them in ancient times, and the living generation has little to say about such gender stories as those from such territories. There are many, however, who still enjoy checking out these old, futile lore that often has hidden mysteries. So, sometimes wondering about this land survey story and finally finding a conclusion that it is worth telling others.

The accounts of the "Icelandic book" (Íslendingabók) and (Landnámu) we know, that the chief author of the first law, which was valid in Iceland, was Úlfljótur, who lived in Lón in East Iceland and the first laws of Iceland were named after him. Now, we don't know, for sure, what was in that law, because they exist no longer as a special book, and not simple to see, what of the Grágáslags-law is from Ulfljot and what is younger. The book Landnáma preserves a few, but remarkable items, which can be considered with complete certainty, existed in the first law. It says inter alia so: "It was the beginning of a pagan law that men should not have a ship's head (figurehead), but if they had, they should take off their heads before they came into a land-sight and would not sail to land with a gaping head and a prow. The fact that this article has been the starting point, or the beginning of the first Icelandic legislation, makes it remarkable that the first thing that they specify is that it is not possible to upset the landguards (country legends). This shows that when Ulfljótslög was established, that belief has been general, that land was guarded with legends, which primarily had the role of guarding them against those who approached the land in a sinister mood. Nationals seem to have looked at the territories as a kind of price - a kind of country and national life guards. From this article Úlfljótslaga, there will be no question of this, what kind of "creatures" these territories were in the court of the ancient. It is first Snorri Sturluson, who says in chapter 33 in the story of Ólafur Tryggvason in Heimskringla. Snorri says that Icelanders were very angry with King Harald Gormsson in Denmark because he let up all the money that Icelandic men had, who had broken their ship in Denmark and called the Danes' Driftage. Then

Icelanders, as so often later, resort to the revenge, which is not the best - nowadays, and Snorri says that it was "in law had in Iceland, that there should be composed a libelous vers about the Danish king, one for each person in Iceland" because of this.

Icelanders setup the land like the tribes of Israel. Symbol of the land was also similar to Ezekiel, except the worm in the east came instead of Lion. There is quite a story concerning this matter and explained in Dagrenning from the year 1946. The idea is to translate the article later on for publishing purpose.

As shown in Thrihyrningur symbol Iceland was likely settled by people using Mathematics, in the future picture will be put into publishing a book concerning this matter. In Iona seems similar system be used and the basic is foot measurement an inch (thumlungur)[11,25,26,27,28)

3

Explaining Pyramidology

Following is various explanation's concerning the GP:

Specialist has viewed their opinions in words, showing clearly their meaning, which is vital for this subject. Therefore, letting their views come here

AR explains:

"Pyramidology is the science which co-ordinates, combines and unifies science and religion, and is thus the meeting place of the two. When the GP is properly understood and universally studied false religions and erroneous scientific theories will alike vanish, and true religions and true science will be demonstrated to be harmonious[1)

Charles Russel's statement concerning the GP.

"This Witness of the Lord in the land of Egypt will bear such testimony as will honour Jehovah (one of God's name) and fully correspond with His written Word. We thus introduce this 'Witness' because the inspiration of its testimony will doubtless be as much disputed as that of the Scriptures, by the prince of darkness, the god of this world and those whom he blinds to the truth" [4]

D. Davidson explains:

"My elucidation of the various phases of the GP's design has led me to perceive that it is an expression of the truth in structural form. I proclaim, with humility and yet with confidence, that the Pyramid's message establishes the Bible as in Inspired word of God, and testifies that Jesus Christ, by HIS DISPLACEMENT, paid the purchase price of mankind's redemption, and effected the salvation of all who truly belief in Him[8]. "Pyramidology is science that deals with the GP's scientific demonstration of Biblical truth, true Christianity and the divine plan respecting humanity on this planet. One who is skilled in this science is therefore defined as a Pyramidology"[1]

Egyptologist who specializes in the study of the GP from archaeological standpoint. Egyptologist does perhaps not know anything about Pyramidology, it is quite necessary for Egyptologist to know about hieroglyphics, but not for the Pyramidology's.

Specialists from Japan tried to build a pyramid, and they discovered difficulties concerning this work. It is still not known how the GP was build, though various hypothesis on the subject.

Why to do research concerning the Bible?

"He that answered a matter before he heard it, it is folly and shame unto him" Proverbs (18:13) Bible[15]

Also:

"Prove all things: hold fast that which is good" I Thessalonians (5:21) Bible[15]

Explanation of AR concerning such research is:

"The Bible reveals that not only has the almighty architect got a plan, but he also has a plan for revealing that plan. Daniel is outstanding Biblical Book of chronological prophecy, and therein the divine decree is the end: many shall travel to and from and knowledge shall be concealed until "the time of the end"- a time when "many shall travel and knowledge shall be increased[1]. THE HUB AND CENTRE OF THE ENTIRE PYRAMID SYMBOLISM IS CHRIST[1].

4

HISTORICAL BACKGROUND OF THE GREAT PYRAMID

GP (GP) is the only of 7 wonders of old time. GP is the only of these monuments still standing.

They were:

- Colossus Iron/brass statue about 30 meters high. (Greece are planning to rebuild)

- Babylon hanging gardens

- The mausoleum Turkey

- Zeus statue Olympia

- Temple of Artemis Greece

- The lighthouse of Alexandria

- GP (GP)

The GP is c.a. 85.000.000 cubic feet.

Every stone and this design have been questioned.

Following text is accurately prepared by AR: (shortened)

"Who builds this wonderful GP? According to Egyptian tradition says "The plans were let down from heaven" or as we would say today, they were divinely inspired[1]".

"The theory that the GP was built by Shem is also incorrect, for we now know that Shem was not living at the time, as his days, on the other hand, were prior to the Pyramids erection; although what available evidence we have indicates that a descendant of Shem, a Semite, was its builder.

"There are direct connections, between the Revelation given to Enoch and that ...enshrined in the GP was Almodad, the son of Joktan (Genesis 10:26) Bible[15].

It is said that the GP is built after a vision from God that Enoch is said to have got about 300 years before the GP was built. Sometimes the GP is called Enoch's pillar, fitting into this explanation. In the Old Testament altar is described not with scarce stones. In New Testament the altar of witness may be with scarce stones.

So, when was the GP built?

This is one of the results from AR, and showing in his pictures, the starting year to build the GP is 2623 B.C.

Without research on every detail of AR idea to be, to say the least questionable. Therefore, wondering how to explain this result of AR.

After a deep study, the result is that AR result is correct. This matter is proven is 5 different ways:

The measurement of the GP is 1 Inch = one year and therefore 2623 B.C.

The historian Herodotus, sometimes called "father of history" confirmed 2622 B.C. As year the GP was under construction.

Three ships owned by Khufu pharaoh were found in the year 1954 and they were examined and confirmed since 2600 B.C. This is a result of C-14 research from 1964 University from Pennsylvania, Philadelphia USA[3].

The Egyptian hieroglyphs confirm that the GP was built when Khufu was a pharaoh in Egypt. This is also confirming the same time frame 2623-2600 B.C. This matter is best explained in the book of AR "Boðskapur pýramídans mikla "from the year 1948.

PS confirms that because of firmament changes during a long period of time slightly, therefore his result was building year was when the Devil star shined down the Descendant passage that fit around 2600 B.C.

Khufu died 7 years after the GP was closed, and therefore the GP was not a tomb. The closing of the GP is complex.

The Chaldaic Paraphrase of Jonathan ascribes to Almodad the credit of being the inventor of Geometry, and likely to be a planer of building the GP.

"Who measured the Earth to its extremities" and as we will see the GP is constructed to marvellous geometric design[1].

Why Egypt? Because Egypt is symbolizing the world.

Peleg was the uncle of Almodad (Gen 10:25-6) and the son of Eber or Heber, whose descendants became known as Hebrews. By the chronological data available we know that the GP was built during the Quarter- century immediately prior to 2600 B.C. It is not impossible for Almodad to have received the plans of the GP under the divine inspiration of course, for God was the architect, Almodad or Elmodad means "God is a friend."

Concerning the building of the GP, there was no slave labour on the work of the GP as most of the other

Pyramids 100.000 men worked in relays for 3 months and had free food. In 10 years, stones were cut, and preparation work is done. The latter 20 years took to build the GP.

Of Peleg was written "In his days was the earth divided" (Gen 10:25 and Chronicles 1:19)[15].

The Lord God hath "measured the waters in the hollow of His hand and enclosed the dust of the earth in a measure and weighed the mountains in scales and the hills in a balance"

"He has weighed the world in the balance, and has measured the times with a measure, and carefully counted the hours, and he will not move or disturb them until the prescribed measure is reached. "(II Esdras 4: 36-37, the Bible[15]).

GP shows 3 different lengths of the year. (Figure 1)

- 365.242 Solar year

- 365.256 Sidereal year

- 365.259 Anomalistic year.

This picture needs further explanation. If you look at this picture then you can see from A to B as Solar year this is the side of the GP, and this has 6 numbers correct for each year in the GP.

Second-year is the Sidereal year, this is on the picture from AEFB, this was not possible to see until the GP was stripped, that is white stones were covering the GP.

Third-year is from (A -b-B), this is also the reason because of violation from Arabs on the GP but showing that the third year is clearly seen. Or anomalistic year.

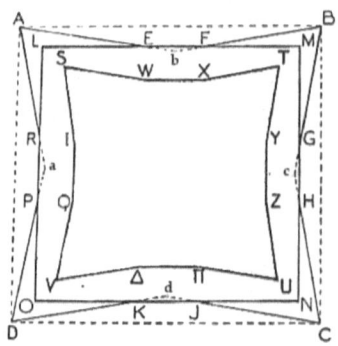

Figure 1: The Great Pyramid ground floor

This picture needs further explanation. If you look at this picture then you can see from A to B as Solar year this is the side of the GP, and this has 6 numbers correct for each year in the GP.

Second year is the Sidereal year, this is on the picture from AEFB, this was not possible to see until the GP was stripped, that is white stones were covering the GP.

Third year is from (A -b- B), this is also the reason because of violation from Arabs on the GP but showing that the third year is clearly seen. Or anomalistic year.

While calculating the possibilities here every year with 6 numbers is in total $(1/10)^{18}$.

"The ancient Arabic Akbar Ezzeman MS., tells us something of its purpose. It states that the Pyramid contains "The wisdom and acquirement in the different arts and sciences... the sciences of arithmetic and geometry, that they might remain as records for the benefit of those who could afterwards comprehend them... the positions of the stars and their cycles; together with the history and chronicle of time past (and) of that which is to come[1)]"

The GP is ideally put to:

- Longitude 31° 9′ 0″ East

- Latitude 29° 58′51″ North

As PS discovered GP standpoint is in the Centre of the Earth[1].

First known to enter the GP was Al Mamoon in the year 820 who made forced passage to enter the GP. He did not find anything of great value, but that was his main aim.

Various persons went to Egypt and admired it's greatness not only of the GP, but also the culture of Egypt. Among visitors of the GP were Napoleon and Colonel Howard Vyse who did measurements and publication of main measurements were clear.

5

HISTORICAL BACKGROUND OF PYRAMID RESEARCH SPECIALISTS: PYRAMIDOLOGY

5.1   John Taylor (1781-1864), his first breakthrough:

John Taylor [JT], the mathematician, published a book in the year 1859 and a breakthrough was made by the English mathematician looking at measurement's from the GP, and his books name was: "The GP: Why Was It Built?"

JT looked at measurements available and discovered that the GP is fulfilling "Squaring of a circle" which means that the base of the GP was the same as circle under the angle 51°51′143″. John's result was that the GP was built under the divine plan. The length of the solar year was in the base of the GP that is 365.242 inch.

As can be seen, in (Figure 2), we have the PI corner of the GP 51°51′143″ under this corner and discovered by JT, "squaring of a circle" mathematical rule is fulfilled in the GP also in (Figure 3). Then in (Figure 1), we can see 3 different years shown in the GP, though JT discovered that the year 365.242 which is one year, and 1 inch is one year. This also means that the Egyptians were not the planer of the GP as they used 365 days in a year, not a fraction 1/4 part of a year.

In this essay, probabilities are calculated in such a form that each number has 1/10 possibilities to come up by chance. 51°51′143″ the corner of the GP by chance is therefore 7 numbers in a row and therefore chance is (1/10)7, also here probabilities of the year with numbers is (1/10)6 and the total chance of being less than (1/10)13.

In this essay probabilities are calculated in such a form that each number has 1/10 possibilities to come up by chance. 51°51′143″ the corner of the GP by chance is therefore 7 numbers in a row and therefore chance is $(1/10)^7$, also here probabilities of the year with numbers is $(1/10)^6$ and total chance of are less than $(1/10)^{13}$.

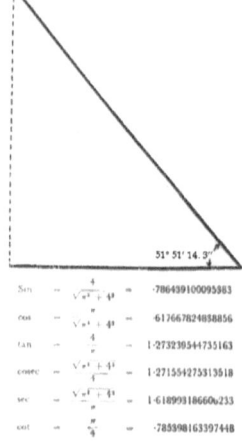

Figure 2: The Pyramidic angle

5.2   Piazzi Smyth 2nd breakthrough

PS was a professor at the University of Edinburg Scotland. He was an astronomer, and that was very useful concerning

GP research. PS was impressed by JT result concerning the GP that is the GP is divinely inspired.

In his research, he discovered that the descendant passage was fitting into Devil star, and because of the time frame, he was finding out when the Devil star was shining down the passage.

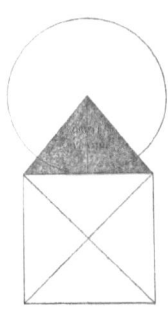

PS result was that GP is in fact "Bible in stone" that is Bible shown in stone, using mathematical information's. AR informs about that Pi is in GP with more exactness than known until 2000 years later. This matter is explained later in this essay.

Figure 3: Geometric Construction of the Great Pyramid

PS agreed to others that 1 inch = 1 year (see Figure 4) and therefore it was both possible to calculate both backward and also forward in the GP. In the year1865, therefore prophecies of the Bible were looked at and the year1914 already under discussions, and prophecies showing difficult times coming.

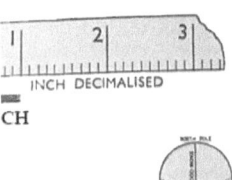

Figure 4 : The key to the scientific revelation of the Great pyramid

PS did not only notice Devil star, but also 7 stars as the throne of God, shining down into the Ante-Chamber, and to be exact to The Boss.

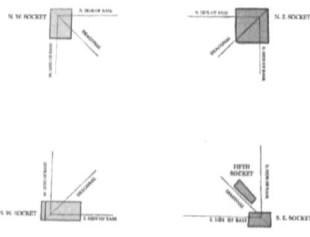

*Figure 5 : The five sockets of the
Great pyramid. Showing their
relative positions.*

In (Figure 5), we see the socket's in the GP and 2 engineers
found 1865 2 of GP socket's. First 2 were found by France
Scientists. Appendix B [2)]

The GP is ideally put to:

- Longitude 31° 9′ 0″ East

- Latitude 29° 58′51″ North

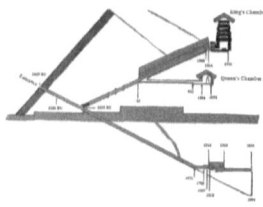

*Figure 6: Hallways and halls
in the Great Pyramid*

In (Figure 6), we can see the score line 2141 B.C. Found by
PS, but originally his measurements were 2140 B.C. These
calculations were confirmed by British specialists in the year

1938. We can measure both back and forth from this point and has not been explained clearly[18].

AR explains, concerning disagreement concerning cubic between PS the astronomer and Sir Flinders Petrie.

Sir Flinders Petrie, sometimes called father of archaeology, he and PS both misunderstood the matter for over 20 years.

The Sacred Cubit and the Royal Cubit (AR)

1 Sacred = Royal Cubits = 1.213204 Royal Cubits
   = 25   Pyramid inches = 25.0266 British inches
1 Royal Cubits = Sacred Cubits = 0.8242637 Sacred Cubits
   = 20.606592 Pyramid inches = 20.62852 British inches

In Figure 7 we see the GP and the Sphinx.

Figure 7

(Figure 8) we see also that the landmass in each 1/4 of the Earth is above sea level similar within demands of scientific demands. This was PS result. PS result was

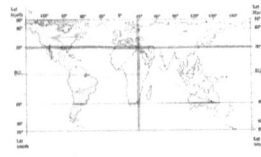

Figure 8:

questioning who knew how the Earth was, way to the moon and sun around 4500 years ago, these massive information's? Beyond any doubt to PS result is that God is behind this building GP, it is divinely, and therefore PS called the GP „Bible in Stone ", that is similar conclusion as other Pyramidology's.

If we add, then if we add (Figure 9 and 10) these two lines goes through the GP. According to each probability is $(1/10)^7$ for each part. If we take two, centre and line $(1/10)^{14}$ probabilities. If all three $(1/10)^{21}$.

Figure 9: Longest line on Dry land goes through the Great Pyramid

These calculations are not included in $(1/10)^{52}$ to calculate this correctly a program must be used. That is to show how small possibilities there is for all these information's. Using part of the information's not all. $(1/10)^{52}$ can be proved, but explaining this is just practical economical thinking.

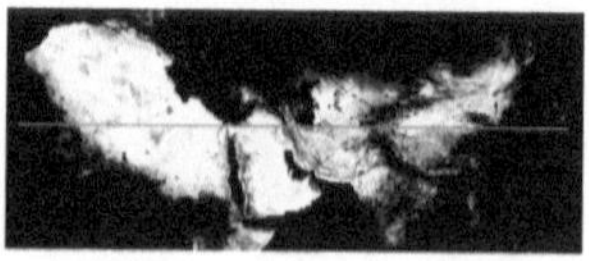

These information's from PS were later useful and explained later in the essay. PS's result was that there is one year shortening

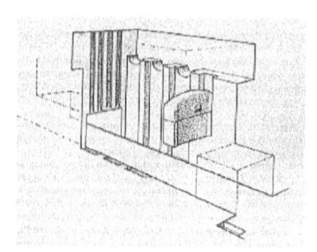

Figure 10: Straight line going through the Great Pyramid

under "The Boss" or sometimes called the "Granit Leaf". (Figure 11) is showing Ante Chamber and the "Boss" which is in the picture like a school bag.

PS and his cooperatives calculated the GP thoroughly, and various new result they found.

PS result was that the Pyramid inch is very similar to British

Figure 11: Anti Chamber

measurement and 1 Pyramid inch = 1,001 British inch and therefore the message of the GP has special meaning for Britain and America who were using the same measurement's, who were more exact than the meter system.

This result of PS that the GP message is mostly to UK and USA. Britain (Ephraim) and America (Manasseh) are surely part of lost tribes of the Israelites, explained in the book of AR: Israel-Britain, or Pyramidology I-IV. AR used part of Israel Britain, or the Icelandic part. Appendix C [5)] Icelandic measurements are also fitting into these British and America measurement's. In Icelandic

Figure 12: Tribes of Israel

faðmur (between hands) = 167 cm and = 31/2 or feet = and foot = 30.48 cm. In Figure 12) we can see symbols for the lost tribes of Israel, and as PS discovered, that the measurement's in the GP has specific meaning to America and Britain as they are using similar measurement's. In (Figure 12) we can see the flags of various tribes, and Ephraim (UK) and Manasseh (USA) and between them is Benjamin (IS) (Figure 13).

While Dr Piazzi was in Egypt, he got a letter from RM:

RM shipbuilder came up with the idea 1865 that perhaps GP fits the Bible prophecy looking at the

Figure 13: Dagrenning

broad way, narrow way and the high way to holiness (King chamber). In the centre of the GP 33 inch would show the lifetime of Jesus Christ who Christian people accept as saviour of the world. RM wrote a letter to PS.

"It was while professor Smyth was in Egypt at the GP that it was revealed to RM, the young Scottish shipbuilder from Edinburgh, during February 1865 that the bust-open upper mouth of the well-shaft represented the death and resurrection of Christ, that the various passages represented the respective ages and that those passages were constructed according to chronological scale of geometric inch to a year. These discoveries of RM gave an impetus to the advancement of Pyramidology. Professor Smyth publicized them by incorporating them in the subsequent editions of his book "Our Inheritance in the GP", for RM did not delay in communicating his discoveries to professor Smyth in Egypt, in a letter dated 25th of February 1865. Professor Smyth's practical work at the GP itself tested and confirmed the findings of JT and RM[4,18].

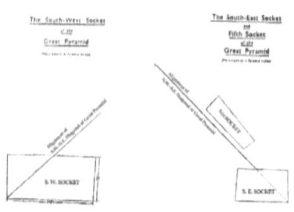

Figure 14:

PS was astronomer of University of Edinburg in Scotland and in the year 1864- 1865 he went to Egypt with Thomas

Inglish and William Aniton they were both engineers from Scotland.

Thomas Inglish and William Aniton found two sockets of the GP in the year 1865. All sockets of the GP are shown in (Figure 14)

- 1801 N-E Le Pére-Colonel Cartelle (France)

- 1801 N-W Le Pére-Colonel Cartelle (France)

- 1865 S- E Aiton – Inglish

- 1865 S-W Aiton – Inglish

- 1925 5th Socket Adam Rutherford Morton Edgar

The result of PS was that he confirmed JT's hypothesis and also Mr. RM ideas. The result of PS concerning Britain and America measurement inch is showing that the message of the GP has special meaning for UK and USA. AR got specialists (around 20-30) and support from authorities for long time for further research.

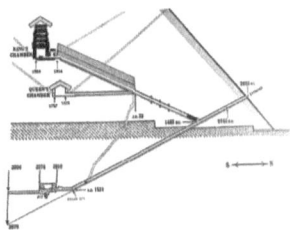

Figure 15: Interior Measuremends of the Great Pyramid

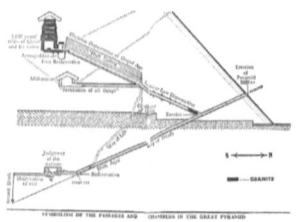

Figure 16:

Figure 16 is showing names given by Pyramidology's in various places in the GP.

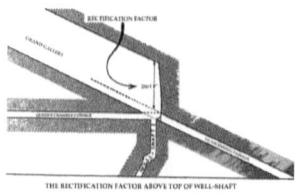

Figure 17: Rectification factor

Figure 17 is showing Christ triangle above well shaft, above the Christ triangle is + 286 rectification factor or payment of displacement factor -286 symbolized for sin of humanity. Or 286-286 = 0 Also Christ triangle angle 26°18′9-7″ goes through birthplace of Jesus Christ.

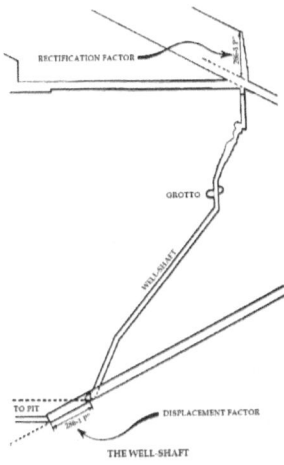

Figure 18: Displacement Factor

Here we see the well shaft showing symbolized new way opened through sacrifice of Jesus Christ from the pit (hell) to heaven.

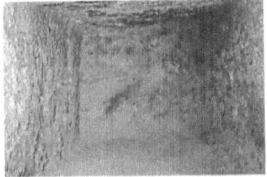

Figure 19: The Pit

5.3 Charles Russell: breakthrough

Charles Russell was a friend of PS, and had suggestions concerning the purpose of the Queen Chamber (The correcting room) and also concerning the King Chamber (the divine way) He publishes his books "The Divine Plan of the Ages" also suggesting that 1914-18 would be a very interesting years to look at from prophecies. He wrote books concerning explaining the text of the Bible. While PS wrote his letter, Charles Russell published the main result of PS. "The Divine Plan of the Ages" (Figure 20)

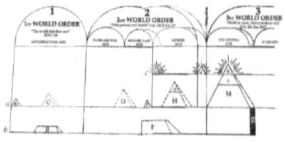

Figure 20: The Divine plan of the ages, portrayed in Pyramid symbols.

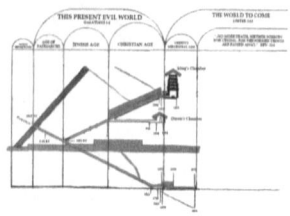

Figure 21: Hallways and halls explained inside The Divine plan of the ages.

5.4 Mr. Robert Menzies, third breakthrough:

RM was a shipbuilder and his hypothesis were that the GP fits into the prophecy of the Bible. Also, 33 inches is fitting Jesus Christ (who is Logos explained by FB) lifetime on Earth. According to RM information in a letter dated to PS, he thought he was of Manasseh tribe. RM suggestion was Ante-Chamber and doorway symbolized the beginning of the final period of Great war and tribulation prophesied in the Bible.

RM also understood "Well of Life" in (Figure 22) we can see the various passage that RM discovered that symbolized prophecy of the Bible.

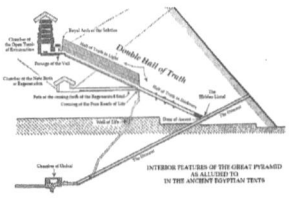

Figure 22: The secret places of the hidden God

In, we can see that is through the sacrifice of Jesus Christ, the GP symbolized crucifixion and opening a new way for humanity. The number 286 symbolized sin. RM understood that people were going down the broad way to the pit. Therefore, Jesus Christ is named "The saviour of mankind"

Menzies theory was adopted by PS, but, unfortunately, he also adopted Menzies idea that the Christian dispensation began at the birth of Christ" vi "GP: "Rev Commander L.G.A. Roberts, who took up the matter with Smyth's followers, but was unable to persuade them to accept his views. About the same time, HA agreed" It was not until 1905 that Col. J. Garnier, R.E. Publishing "The GP, Its Builder and Prophecy" This was corrected, that is Christian dispensation began after the crucifixion of Jesus Christ, not his birth[8].

5.5 Dr Flinders Petrie (1855-1942) next Pyramid specialist

Dr Flinders Petrie (1855-1942) was an atheist and is sometimes called the father of archaeology. He measured the GP very carefully, and brought up more exact measurements, from the GP. There came up a disagreement with PS concerning the length of the cubic. That debate delayed publishing about the matter for 20 years.

AR's conclusion was that both PS and Dr Flinders Petrie misunderstood in the cubic debate that both measurements were linked to each other. Also, AR proves the inch. Showing the complexity of the matter. AR is explaining that the measurements of PS and FP are related, and in fact confirming each other.

5.6   Dr John Edgar and his brother Morton Edgar

Dr John Edgar and his brother Morton Edgar did careful measurements and calculation's especially concerning shafts of the GP, also they discovered the corner of Christ birthplace Bethlehem (Figure 23)

Were very exact in their measurement's and managed to measure other places others could not do.

GP Passages and chambers published 1923- 4. The finding of the corner of

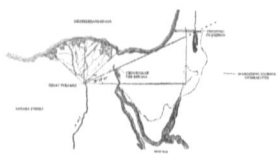

Figure 23: Christ corner.

Wandering journey of Israelites

Christ or 26°18′9-7″, confirmed later on. By (AR) and others.

These pictures show Christ angle, going through Bethlehem. Showing accurate date of Jesus Christ life on Earth with mathematics. AR has proven this as can be seen in various pictures mentioned here. First to confirm these findings of Dr John Edgar and Morton Edgar were, Captain John Mace ague and William Orr Warden (DD/HA) in the introduction in their book[8].

Christ angle and day's he lived on Earth.

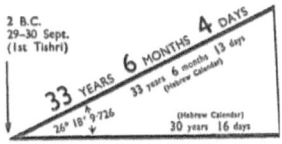

Figure 24: Christ angle.

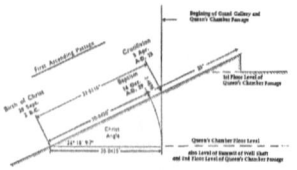

Figure 25: Christ triangle.

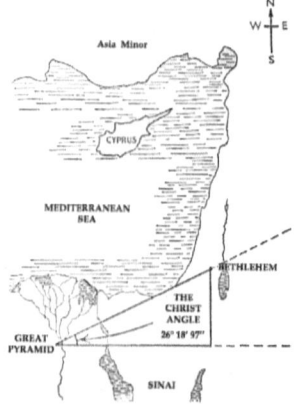

Figure 26: Map of Christ Angle.

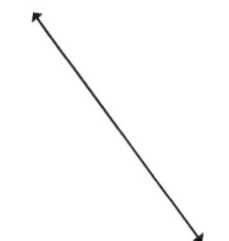

Exodus: Moses crosses (D.D.)

Following proofs, Figure 27-31, are made by AR[1-4]

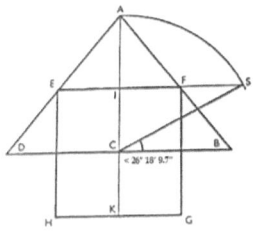

Figure 27: Geometric construction of the Christ Angle.

PROOF: (F27)

BD = y Sacred Cubits $\therefore$ AC = $\frac{2y}{\pi}$ Sacred Cubits.

Area of $\triangle$ABD = $\frac{y}{2} * \frac{2y}{\pi} = \frac{y^2}{\pi}$ square Sacred Cubits

$\therefore$ EF = $\sqrt{\frac{y^2}{\pi}} = \frac{y}{\sqrt{\pi}}$; IC = $\frac{EF}{2} = \frac{y}{2\sqrt{\pi}}$; CS = $\frac{2y}{\pi}$ Cubits

$\therefore$ sine $\angle$ ISC = $\frac{IC}{CS} = \frac{y}{2\sqrt{\pi}} + \frac{2y}{\pi} = \frac{y}{2\sqrt{\pi}} * \frac{\pi}{2y} = \frac{\sqrt{\pi}}{4}$

$$\angle BCS = \angle ISC$$

$\therefore$ sine $\angle$ BCS = $\frac{\sqrt{\pi}}{4}$ = 0.443113462726

$$= \frac{\pi y}{4\sqrt{\pi}*y} = \frac{\sqrt{\pi}}{4} = 0.443113462726$$

Which is = sine of the Christ Angle,
26° 18' 9.7" Q.E.D

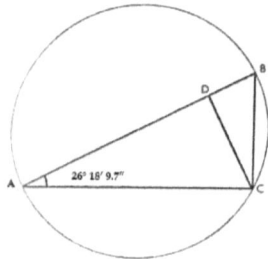

Figure 28: Geometric
construction of the Christ
Angle. Alterntive Method

PROOF: (F.28)

$AB = \frac{2y}{\pi}$ Sacred Cubits: $BD = \frac{y}{8}$ Sacred Cubits

$\angle$ ABC is the complement of

$\qquad \angle$ BAC, complement of $\angle$ BCD

$\therefore \angle$ BAC $= \angle$ BCD

Sine $\angle$ BAC $= \frac{BD}{BC}$; sine $\angle$ BAC $= \frac{BC}{AB}$

$\therefore \frac{BD}{BC} = \frac{BC}{AB}$ $\quad \therefore (BC)^2 = BD * AB$

$(BC)^2 = \frac{y}{8} * \frac{2y}{\pi} = \frac{y^2}{4\pi}$

$\therefore BC = \sqrt{\frac{y^2}{4\pi}} = \frac{y}{2\sqrt{\pi}}$

Sine $\angle$ BAC $= \frac{BC}{AB} = \frac{y}{2\sqrt{\pi}} + \frac{2y}{\pi}$

$\qquad = \frac{\pi y}{4\sqrt{\pi * y}} = \frac{\sqrt{\pi}}{4} = 0.443113462726$

$=$ sine of the Christ Angle, 26°18′9.7″

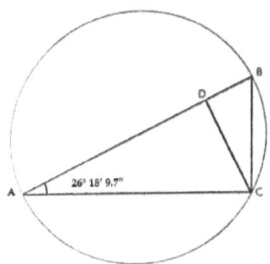

Figure 29: Mathematical representation. Symbolic Factors.

PROOF: (F29)

Base side of Gt. Pyramid = y Sacred Cubits

$\qquad = 5^2 y$ Pyr. inches

Semi-diagonal $= \frac{5^2\sqrt{2}}{2}*y$ Pyr. inches

$\therefore \frac{1}{10}$ Semi-diagonal $= \frac{5^2\sqrt{2}}{2*10}*y = \frac{5\sqrt{2}*y}{4}$ Pyr. inches

$\therefore BC = \frac{5\sqrt{2}*y}{4}$ Pyr. inches

$\qquad = \frac{5\sqrt{2}*y}{4} * \frac{\sqrt{\pi}}{4} = \frac{5\sqrt{2\pi}*y}{2^4}$ Pyr. inches

- Rectification Factor $= \frac{5\sqrt{2\pi}*y}{2^4} = 286,10213 \; Pyr.\,inches$

- Displacement Factor $= -\frac{5\sqrt{2\pi}*y}{2^4} = -286,10213 \; Pyr.\,inches$

- Expansion Factor $= \frac{5\sqrt{2\pi}*y}{2^7} = 35,76277 \; Pyr.\,inches$

- Contraction Factor $= -\frac{5\sqrt{2\pi}*y}{2^7} = -35,76277 \; Pyr.\,inches$

Q.E.D

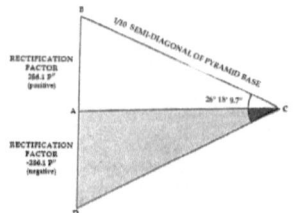

Figure 30: Christ Angle. Royal Cubit Method.

PROOF: (F30)

$$AB = y \text{ Sacred Cubits} = \frac{10^3\sqrt{\pi}*y}{4y} \text{ Royal Cubits}$$

$$= \frac{10^3\sqrt{\pi}}{4} \text{ Royal Cubits}$$

$$\text{Sine } \angle ACB = \frac{AB}{AC} = \frac{10^3\sqrt{\pi}}{4} / 1000$$

$$= \frac{10^3\sqrt{\pi}}{4*10^3} = \frac{\sqrt{\pi}}{4}$$

$$= 0,443113462726$$

= sine of the Christ angle, 26°18'9.7"

Q.E.D

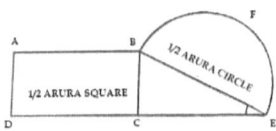

Figure 31: Christ Angle-Arura Method.

PROOF: (F31)

Area of Arura = 10^4 square Royal Cubits

$\therefore$ Diameter BE = $2\sqrt{\dfrac{10^4}{\pi}}$ Royal Cubits

$$= \frac{2*10^2}{\sqrt{\pi}} \text{ Royal Cubits}$$

Sine $\angle$BEC $= \dfrac{BC}{BE} = \dfrac{50}{\frac{2*10^4}{\sqrt{\pi}}}$

$$= \frac{50\sqrt{\pi}}{2*10^2} = \frac{\sqrt{\pi}}{4} = 0{,}443113462726$$

= sine of the Christ angle, 26°18'9.7"

5.7 Dr Herbert Aldersmith and David Davidson were next pyramid specialists

HA who was a physician his interests were to specialize in prophecy and was in agreement with PS in most cases, except well of life is measured concerning crucifixion, not the birth of Christ. HA suggestion was that the year 1928 would be of special interest concerning the End time. His cooperative was DD structural engineer who was an atheist but changed his views when doing more research on the subject. After the death of HA in the year 1918, DD carried on with the work, and J. Clemishaw went over the matter and corrected errors made. Later on, DD did calculations concerning Ante-Chamber, this can be seen on the picture DD 1921. No conclusions from the update were made, after recalculated Ante-Chamber. This had vital importance for the result, as shown later in Ph.D. Essay.

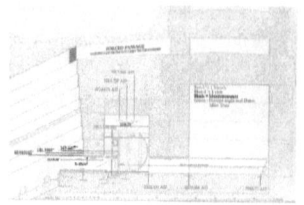

Figure 32: Ante-Chamber.

5.8 Adam Rutherford, his son James Rutherford

AR was outstanding in his work and discovered many new things and also looked over the work of former Pyramid specialists. Concerning information's in Pyramidology Book I-IV after AR and his son James Rutherford's, there are no other errors beside the End Time calculations.

It was to their great disappointment that the End Time calculations were wrong. The last group of specialists' and researchers were AR, James Rutherford. AR was introduced to the Pyramidology through his father. James Rutherford (older).

AR discovered through the 5th socket, that Icelanders were between the scored lines, and therefore he believed that the tribe of Benjamin is God's chosen. James Rutherford carried on with his father work. They both died in car accidents. Top stone has under it 286 which is symbolized by sin.

In Table 1 is information that has not been explained before, are calculations by James Rutherford (AR father). According to Jonas Guðmundsson Vördubrot O. De. Blaere from Antwerpen Belgium had also discovered the following:

Table 1: The Great Pyramid text of scripture, as arithmetical figures. Value of every letter in the Hebrew bible is given and each line represents a word. Isaiah 19: 19-20

											=	
2	+	10	+	6	+	40	+	0	+	0	=	58
5	+	5	+	6	+	1	+	0	+	0	=	17
10	+	5	+	10	+	5	+	0	+	0	=	30
40	+	7	+	2	+	8	+	0	+	0	=	57
30	+	10	+	5	+	6	+	5	+	0	=	56
2	+	400	+	6	+	20	+	0	+	0	=	428
1	+	200	+	90	+	0	+	0	+	0	=	291
40	+	90	+	200	+	10	+	40	+	0	=	380
6	+	40	+	90	+	2	+	5	+	0	=	143
1	+	90	+	30	+	0	+	0	+	0	=	121
3	+	2	+	6	+	30	+	5	+	0	=	46
30	+	10	+	5	+	6	+	5	+	0	=	56
6	+	5	+	10	+	5	+	0	+	0	=	26
30	+	1	+	6	+	400	+	0	+	0	=	437
6	+	30	+	70	+	4	+	0	+	0	=	110
30	+	10	+	5	+	6	+	5	+	0	=	56
90	+	2	+	1	+	6	+	400	+	0	=	499
2	+	14	+	200	+	90	+	0	+	0	=	293
40	+	90	+	200	+	10	+	40	+	0	=	380
20	+	10	+	0	+	0	+	0	+	0	=	30
10	+	90	+	70	+	100	+	6	+	0	=	276
1	+	30	+	0	+	0	+	0	+	0	=	31
10	+	5	+	6	+	5	+	0	+	0	=	26
40	+	80	+	50	+	10	+	0	+	0	=	180
30	+	8	+	90	+	10	+	40	+	0	=	178
6	+	10	+	300	+	30	+	8	+	0	=	354
30	+	5	+	40	+	0	+	0	+	0	=	75
40	+	6	+	300	+	10	+	70	+	0	=	426
6	+	200	+	2	+	0	+	0	+	0	=	208
6	+	5	+	90	+	10	+	30	+	40	=	181

5,449

Table 2 is showing King Chamber numbers.

Table 2: The Great Pyramid mathematical values. Correct to 5 decimal places.

Description	Geometrically Measurement*	Determined	Formula
Length (E-W)	412.13186 Pyr. Inches		
Breadth (N-S)	206.06593 Pyr. Inches		
Height	230.38871 Pyr. Inches		
Diagonal of Floor	460.77743 Pyr. Inches		
Diagonal of End walls	309.09889 Pyr. Inches		
Diagonal of Side walls	472.15636 Pyr. Inches		
Cubic Diagonal	515.16482 Pyr. Inches		

Table 3: The Great Pyramid mathematical values. Correct to 5 decimal places.

Description	Measurements
Pyramid π < (Gt. Pyramid's faces	51° 51' 14.30645"
Arris < (Gr. Pyramid's corner)	41° 59' 50.00622"
Christ < (Angle of main passages)	29° 18' 9.72609"
Quadrature <	27° 35' 49.60549"
Base side length-full design	365.24235 Sacred Cubits
Base side length-as constructed	362.38133 Sacred Cubits
Height of Pyramid-full design	232.52050 Sacred Cubits

Height of Pyramid to summit platform as built	217.94944 Sacred Cubits
Displacement Factor	286.10213 P. Inches
Rectification Factor	286.10213 P. Inches
Contraction Factor	35.76277 P. Inches
Expansion Factor	35.76277 P. Inches
Sidereal Year-Pyramid value	365.25636 m.s.days
Solar Tropical Year-Pyramid value	365.24235 m.s.days
Anomalistic Year-Pyramid value	365.25986 m.s.days

(Figure 33 showing Nile delta quadrant)

Henry Michael who was CEO over Navy's drawings and when he looked at the river Nile that was 1/4 of a circle, he wanted to see what the centre of the circle was. When he discovered that it was the GP, then he said: "This building is the most important in the world[1]"

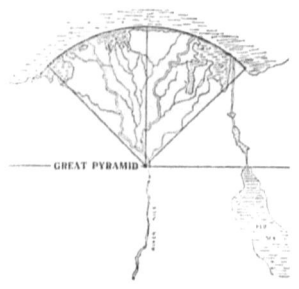

Figure 33: The Nile delta quadrant.

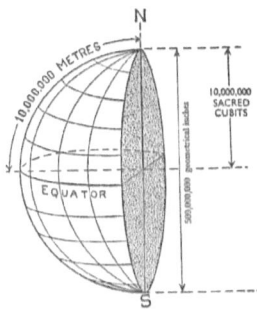

Figure 34: Ancient sacred cubit-more accurate than the mordern metre.

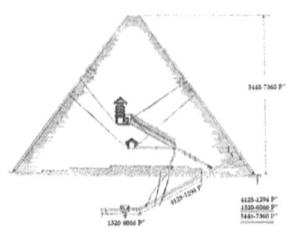

Figure 35(13): 5449th inch in the design of the Great Pyramid.

Out of Israel's 12 tribes, Joseph and Benjamin had the same mother Rachel and father Jacob. All the other were the half-brother of Joseph, who cheated Joseph who ended up in Egypt, while Benjamin did not take any part in that. Joseph went to Egypt and had a wife from Egypt, and they had two sons Ephraim and Manasseh.

Because of complexity, AR was hesitating though concerning the matter, even though he found the fifth socket with Morton Edgar.

Main work of AR, was Israel Britain showing that the GP was linked to Britain and America. The Pyramid inch is 1 inch = 1 year = 1,001 British Inch. PS who calculated the inch and he was sure that connections between Britain and America as their measurements are in line with GP measurements. This seems to be correct. Old Icelandic measurement's foot etc. was similar to UK and USA measurements.

AR had discovered the 5th socket and showing Iceland between the main lines.

This was discovered in the year 1925 by AR and Morton Edgar.

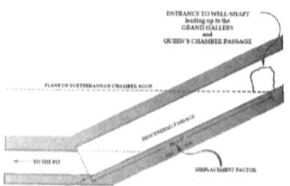

Figure 36: The displacement factor below bottom of Well-Shaft.

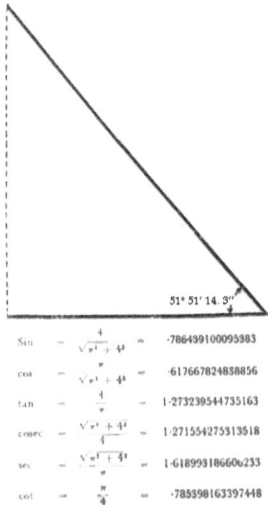

$\sin \quad = \quad \dfrac{4}{\sqrt{\pi^2 + 4^2}} \quad = \quad \cdot 786439100095883$

$\cos \quad = \quad \dfrac{\pi}{\sqrt{\pi^2 + 4^2}} \quad = \quad \cdot 617667824838856$

$\tan \quad = \quad \dfrac{4}{\pi} \quad = \quad 1 \cdot 273239544735163$

$\mathrm{cosec} \quad = \quad \dfrac{\sqrt{\pi^2 + 4^2}}{4} \quad = \quad 1 \cdot 271554275313518$

$\sec \quad = \quad \dfrac{\sqrt{\pi^2 + 4^2}}{\pi} \quad = \quad 1 \cdot 618993186606233$

$\cot \quad = \quad \dfrac{\pi}{4} \quad = \quad \cdot 785398163397448$

Figure 37: The pyramidic angle.

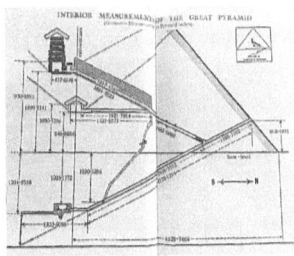

Figure 38: Inside of the GP. Queen's Chamber measurments (1521 Luther church established that year, and (1727) death year of Sir Isaac Newton.

Newton End Time year 2060 is in the GP 1030.3296 x 2 = 2060, 6592

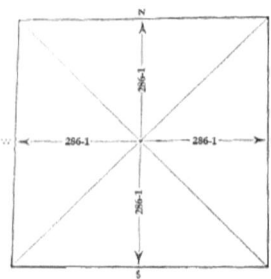

Figure 39: Rectification factor
(286) in the Top-Stone.

Now, A.D. 1521 is clearly defined in world history as the beginning of a new epoch. That year witnessed the beginning of the Great reformation and the close of the Dark ages. On April 17th the famous Diet of Worms was held and on May 26th, 1521 the great Reformer Martin Luther was officially outlawed, whereupon Europe was split into Roman Catholicism and Protestantism[1].

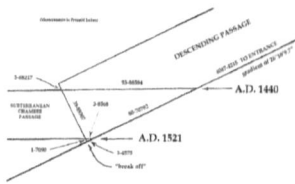

Figure 40: Beginning of
subterranean chamber passage.

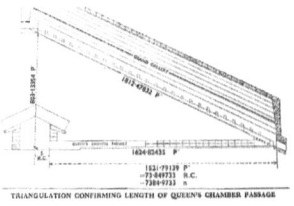

Figure 41: Triangulation
confirming length of Queen's
Chamber passage. 1521 is
symbolic for Luther, and longer
line is 1727 death year of Sir
Isaac Newton.

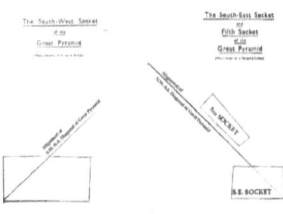

Figure 42: Fifth socket.

Figure 43: Íslandsrákin.
Icelandic beam.

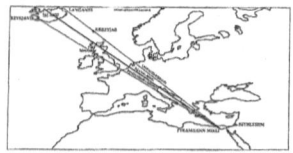

Figure 44: Íslands Ásinn.,
Icelandic beam.

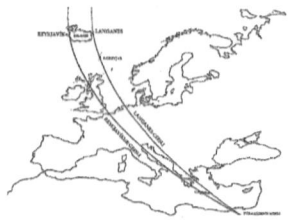

Figure 45: Íslandsrákin.
Icelandic beam.

Iceland is between the lines and discovered by AR from 5th socket Israel-Britain AR, Pyramidology I-IV by AR and James Rutherford engineer, and Iceland great inheritance AR.)[6]

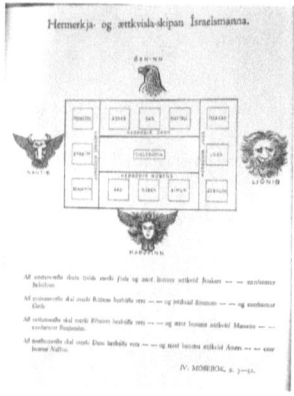

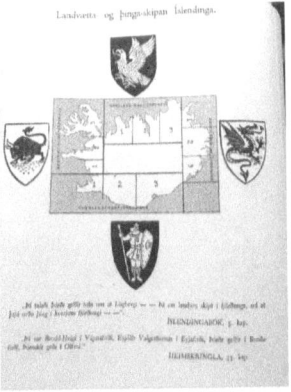

5.9   Fred Binns ex-president of Institute of Pyramidology, 2019, from UK.

Fred took over the Institute of Pyramidology and published Pyramid quarterly. Fred is a big reader and his main interest is Bible and he knows of course of the Pyramidology. Fred did publish Pyramid quarterly and brought many new views to the subject. Fred carried on the work of AR and James Rutherford who was master in engineering. Main work and developing the research was AR. Not only developing this matter further but also doing thorough research on former writers on the subject. This essay is simplifying matters and because of complexity one must specify, and then bring the main theory developed by these specialists. Appendix C you can see letter from FB.

6

MAIN RESULT OF PYRAMID SPECIALISTS

It is suggested that the GP is a repository of ancient knowledge in which it is demonstrable that an advanced level of mathematics was known including Newtonian mechanics.

6.1 David Davidson structural engineer

- The following is a list of conclusions drawn from the analysis of the relationship of dimensions found within and without the GP:

- A precise definition of the Royal Cubit as it relates to the Earth

- The size and shape of the Earth - The Mass and Density of the Earth

- The Gravitational Constant - The Escape Velocity from the Earth to obtain an Open Orbit

- The Escape Velocity from the Earth to obtain escape from the combined Earth's and Sun's gravitational field

- The significance of the location of the GP

- The Golden Ratio

- The Mass of the Sun

- The Mass of the Moon

- The Mean distance to the Sun and the Circumference of the Earth's Orbit

- Neutral Points of Gravity between the Earth and the Sun

- The Mean distance to the Moon

- The Orbital Velocity of the Earth

- The Orbital Velocity of the Moon

- The Metonic19 year cycle of the Moon's orbit of the Earth

- The Lagrange Point (L1) between the Earth and the Moon

- The Speed of Light

- The Orbital Velocity of the Solar System relative to the Centre of the Milky Way Galaxy

- The Velocity of the Local Group of Galaxies which includes the Milky Way Galaxy relative to the Universe. References [8]

6.2 Adam Rutherford and James Rutherford

The spheroidal shape of the earth the amount of to which the earth deviates from the true sphere owing to what is commonly known as "flattening of the Poles"

- The precision of the equinoxes

- The length of Earth apex of rotation

- The duration of the Earth means solar tropical year (equal to the sun appears to make a complete revolution in the ecliptic taken from one vernal equinox to the next 365.242 days

- The length of the Earth sidereal year the actual time that the year takes to revolve around the Sun measured by stellar observations i.e. 365,256 days

- The period to make a complete revolution in its orbit from periheim to periheim i.e. 365, 259 days

- The length to the Earth's orbit

- The mean distance of the Earth to the Sun

- The volume of the Earth crust above mean sea level

- The proportion of land and sea on the surface of the Earth

- The weight of the Earth

- The obliquity of the Ecliptic

The GP is the most accurately oriented edifice in the world its four sides face due North, south, east and West respectively (the entrance is on the north face of the building).

It is built at the centre governing are of the quadrant of the Nile Delta.

It is built at the geographical centre of the land surface of the Earth (refer to the map of the World on a homolographic projection.

It stands on the longest land contact meridian on the Earth surface.

It is situated on the longest contact Earth circuit bearing (chump) on the Earth surface.

Pyramid Quarterly FB and similar to AR Pyramidology I-IV[19]

6.3   Erlingur Thorsteinsson concerning the Great
      Pyramid and information's confirmed

- Mean distance to the Sun

- Speed of Light situation of GP North

- Centre of the Earth-circuit PS.

- The longest line on land goes through the GP

- 3 various years are in the GP (GP)

- Law of gravitation is in the GP. Useful for Sir Isaac
  Newton

- Squaring of a circle JT confirmed by PS.

- Pi is more accurate in G.P. than 2000 years later AR

- Well of Life discovered by R. Menzies and confirmed
  by me and discovered also 153 years later.

- 5th Socket, AR and Morton Edgar showing
  Benjamin tribe in Iceland

- Christ angle discovered by Dr John, and Morton
  Edgar and confirmed goes through Bethlehem.

- Prophecies in the Bible GP and confirmed by PS

- Ante-Chamber DD and recalculated reinvented
  by ET

- Ante-Chamber 93 years of delay 1844-2051 ET

- Granit Leaf and Boss finding ET and USA Embassy 14 May 2018

- End Time calculations ET and Sir Isaac Newton

- Benjamin blood tracing by SB

- Adam tracking by SB and AR

- Isaiah 19 19-20 J. Rutherford father of AR 5449 calculated 5449 Inch height of G.P.

- Rom 11.1 final proof of Benjamin tribe not rejected. ET and confirmed by FB.

- The entrance year 2623 B.C found by AR and confirmed by ET.

- Queen Chamber reinvented 1521 and 1727 explaining matters fully. ET

- Going over matter yes by reading, but further confirmation's come after doing research concerning the building of GP.

- J. Clemishaw in the book "The GP Its Divine message" in the introduction, that the work of DD naturally divides itself into following subjects.

- The History of geometry and metrology

- Gravitational astronomy

- Astronomical chronology

- Archology and history & theology

- J. Clemishaw: This former civilization had condensed its knowledge of natural law into a single general formula, and the application of this formula was analogous to the modern application of Einstein's theory of relativity[8].

(This means that the Einstein's theory is in the GP around 4600 years before Einstein discovered) The GP its Divine message DD. HA 1925 updated later)

These comments of J. Clemishaw needs further investigating and outside of the scope of this essay.

7

Pyramid researchers

- GP has been the world wonder and its stunning accuracy. It is mentioned by Plato, and Pythagoras, and when Jesus Christ was living GP was c.a. 2600 years old.

- To have this here and from Pyramidology view's advance, these have brought this matter where we are today. According to new information's F. Norton is mentioning four Pyramids and the fourth is called black Pyramid.

- For further readings and information's Internet.

- 737 Fredric Norton (Denmark)

- 1765 Nathaniel Davidson

- 1798 Napoleon and research team Jomard and Coutelle

- 1830 Colonel Howard Vyse

- 1859 JT mathematical calculations

- 1865 PS doctor astronomer and researcher concerning the GP

- 1865 Charles Russel

- 1880 William Flinders Petrie professor

- 1900 Dr John Edgar and his brother Morton Edgar

- 1900- 1956 H Aldersmith Doctor in Physics (died 1918) and Structural Engineer D Davidson

- 1900-1990 AR and his son, Engineer, James Rutherford

- 1990-2016 FB inventive writing as president of Institute of Pyramidology UK.

- 1992-2019 Erlingur Thorsteinsson End Time calculations Economist/teacher and inventor, researcher.

8

Various information

Various specialists concerning the GP. This is just part of the specialist's, and information's concerning the subject.

Alan Alford: a basic explanation of Egypt religion's not connecting so much to Christianity.

Ryszard Kaminski: still calculating mostly from an engineering point of view perhaps finding the 4th length of the year among other findings.

Graham Hancock: Various books concerning the subject.

Samuel Laboy: Engineering and recalculated subject concerning GP. Appendix C.

L. Pahl: American Institute of Pyramidology (AIP).

Ancient architects: Video's interesting researcher in various fields.

Broadway passage has measurement 1521 Inch which is symbolized by Luther reformation. Also, up to the narrow way 1521 formal split year between Catholic Church and Luther Church. In the Queen chamber which is explained as correction room, 1521 Inch Luther Church is also there. This was discovered 11 February (2018) as can be seen also 1727 (the death year of Newton) is in that room longer line and explanation is done later in this essay. No religion is above critical view's.

According Avi Lipkin staying here (9-11 May 2018) same problem is in Judaism, no one is 100% correct, so same problem as with the churches.

- Sunday   Sunnudagur (Day of the Sun)
- Monday   Mánudagur (Day of the Moon)
- Tuesday   Þriðjudagur (means 3. day)
- Wednesday Miðvikudagur (means middle of week)
- Thursday   Fimmtudagur (means 5. day)
- Friday   Föstudagur (Fasting day)
- Saturday   Laugardagur (Bathing day)

Various text in the Bible seems fitting into GP.

"Sir Gaston Maspéro, discoverer of the Pyramid Texts, declared "The Pyramids and the Book of the Dead reproduce the same original, the one in words, the other in stone" Whilst that statement is partly imagination and exaggeration it is not entirely so for underneath the corrupt religion of that time there was a heavy substratum of truth and it is this which finds its counterpart in GP symbolism and the true position can be expressed in the statement: The GP and the Bible reproduce the same original, the one in words, the other in stone. The elucidation in detail is given in Book III, pages (1097-1101) and a chart showing the names of the passages and chambers in the GP as alluded to in the Book of the Dead is given on pages 1188-1189 and a summary of these names tabulated on[4].

Egyptologist W. Marsham Adams believed, as many still do today, that the GP was a Temple of Initiation into the mysteries and that the Pyramid Texts and papyri of the Book of the Dead describe that initiation as it was enacted in the GP. It must be realized however that even if the ritual did pertain to ceremonies involving the GP's passages and chambers, it was not necessary actually to go to the Pyramid. We have a parallel of a similar kind in the Christian religion[4].

Who built the GP? "The forefathers of Hebrew who were the descendants of Shem, and this is confirmed by Josef the Jewish historian, after a vision 300 years before the great flood of Noah. The GP was sealed 7 years before Khufu died, his tomb was not in the GP[7].

The GP is in the centre of the Earth-circuit. What most people do not have a clue about is that the landmass in each quarter is even. This means that the author had knowledge about the Earth. Long before other persons on the planet. One can say that is merely a chance. The info is so detailed that the builder knew, not only about the Earth but also the distance to the moon and sun. This was finding of PS. He also calculated 1 inch = 1 year. Pyramid inch and British

Inch is very similarity 1 to 1.001 the latter number is fitting to 286 or sin. PS and Sir Flinders Petrie had a debate 20 years concerning the cubic and misunderstood both that Scared Cubit (25.0266) and Royal Cubit (20.6285) are related.

What people seem not to know is that Mathematics we have certain rules concerning possibilities, the question is about how likely certain matter is to happen. When doing measurements, for example in AR's books he uses 3-4 numbers after the comma, which means the exactness of the measurements. That would in most cases be acceptable depending on what kind of research is made.

The GP is at the border and also fulfilling prophecies of being at the border and mid of Egypt. To explain this matter further, just concerning the picture that the GP is at the centre of the Earth, and each quarter has equal landmass area if we look at this matter from pure chance possibilities the chances are there, and the chance are so tiny fragment that it is agreeable even by scientific point of view that this is almost impossible. If we add to that that there is a line on longest land going directly through the GP, the likelihood is there but within all reasonable doubt that the author knew about the Earth design (this is PS result).

If we add to that knowledge of Pi with exactness not possible to know many thousands of years before, say AR's calculations correct, then 4600 years ago it seems known to the author.

From the chance point of view plus knowledge of the distance to the Sun and the Moon, and also showing at least 3 different lengths of days, and also if 1 = Inch = 1year measurements are used we have too many correct dates in this matter

9

New findings concerning End Time calculations and research work

Research started with the book "Iceland great inheritance" by AR and some other books by Sigfus Eliasson who is mentioning "Pyramid of truth", while in the Pyramid books this AR is mentioning 286 as an error in the GP.

This seems not to fit. If the GP was "Pyramid of truth" then there could not be such obvious error, in the GP. This was in fact the key understanding concerning this matter.

7.1   End Time hypothesis 1993-4 returned to Fred Binns, and confirmation's.

This matter is very specialized and complex.

7 x 286 = 2002 as 286 is 11 x 13 x 2 as sin.

Then 286 + 2 = 288 = (some explain as note D) (12 x 12 x 2) multiplied with 7 is 2016.

Therefore, ET calculated End time was 2002-2016-17

Adding that 2002-2016/7 has about 10 years shortening, this is discovered by insight, but it can be explained that 5 is grace and while a person comes here to say "the good news" again so it is 2 x 5 or 10 years in shortening. Number 10 is explained as a perfect number.

That is something drastic would happen 2002-2006/7.

In the year 2008, there came Bank fall and discovering that the Bank fall happened in fact 2006 and was delayed for 2 years. Here 3 / 10 of the biggest bank fall were in Iceland, and we are only about 300.000 people.

In the year 2015, FB said that "their person" had done calculations concerning the year 2015. Looking at the matter and discovering that Sir Isaac Newton had calculated 23 of September 2015. What happened then? The Catholic Church gained more power through parliament of USA and the United Nations on the mentioned day. Newton had no doubt God exists. Appendix C [3]

Then 2 years later that is 23 September 2017 there was a prophecy fulfilled in the Sky Revelation 12. Virgin birth took place and the whole matter has been misunderstood according to the view. End Time sign[21].

Also 4 blood moons were seen in the Sky, and on the feast days of the Jews[23].

Figure 46: Four Blood Moons

Then Fred arrived in Iceland in the end of April and stayed here for about seven days. We had a Pyramid conference in Grensás church 1 May 2016. FB only said about the Pyramid message, but said that we could do video's and message concerning AR's result concerning Iceland. FB knew about the book, but later on discovered that he was not so keen concerning that result.

Then 2 years later that is 23 September 2017 there was a prophecy fulfilled in the Sky Revelation 12. Virgin birth took place and the whole matter has been misunderstood according to the view. End Time sign[21].

Also, 4 blood moons were seen in the Sky, and on the feast days of the Jews[23].

This you can find out yourself, concerning the meaning[22].

Then at the beginning of the year 2016, discovering about the Ante-Chamber picture signed by D.D from the year 1921. In the Ante-Chamber in centre of circle 2016,257 which is 3-4 April 2016. At the beginning of 2016, showing both Fred and the Pyramid group that perhaps something drastic would happen these days or 3-4 of April. Panama papers, and later the prime minister had to resign. This was a global finance scandal.

It is likely that FB had not thought so much about the calculations looking at the specialist's calculating up to date into the future, which is no easy task. The calculations were only years.

The famous economist John Maynard Keynes had bought Newton's papers[9]. After that surely in the year 2017 noticing that Newton had done some calculations before.

7.2 Dr Bragi Árnason professor in Chemistry at University of Iceland

"When 2 persons, calculate with 2 different methods the same matter and get the same result, you cannot get a better proof" (This is a general rule concerning science according to Bragi Árnason who did not investigate Pyramid matters especially)

Figure 47: Stone of Destiny, under the seat.

• Sir Isaac Newton

    2000-2016-2050/2060

• ET 2002-2016/7-2046/2051

On a visit to Britain in November 2017, where FB informed about that "their person" did these calculations first. Newton wrote a book about Revelation of the Bible and the book of Daniel. Newton's result was the older papers of the Bible the more correct. Behind the Bible is Mathematics.

In Westminster Abbey is a burial of most of Britain's high-rank people. Among them are, for example, Winston Churchill, former prime Minister of the United Kingdom, and the most important person against Hitler in WWII. Churchill had a few very disturbing decisions to take, to secure Britain he had to attack the France fleet, because of the danger that the Germans would use the ships to attack Britain.

Winston Churchill grave is in another place, as he said it is enough to be walked over during life time.

Concerning Britain leaving The Europe Union (EU), it is while visiting Westminster Abbey and meeting a person outside in a rather dreadful situation. Asking if this is not the Parliament of UK, and the answer was yes. This was Britain main fighter for leaving EU. The view was that UK never got anything fighting mostly alone against Hitler. America came helping in correct moment.

Churchill understood the meaning of Stone of destiny under the British throne.

End Time calculations: What do I mean by that?

The meaning is that the World system will be changed, and we will have a better system taken up globally, where the Law of God is followed and his book, The Bible.

To name a few doing these calculations and opinions.

PS was wondering about 1881 looking at Grand gallery, and 1914 in the year 1865.

Charles Russel thought about 1881 and that 1914-1918 the change would come.

HA idea was 1928 would be End Time significance

DD, and AR and James Rutherford calculated 1953

AR and his son James Rutherford re calculated 1979.

ET 2002-2051 (first to 2016/7 and later to 2051 after Ante-Chamber information)

Sir Isaac Newton 2000-2060 (or 2016-50)

7.3   The Ante-Chamber

In the Ante-Chamber seems to be 93 years delay of King Chamber. Explaining this you can look at King Chamber side 1928 – 1953 or 25 years (Figure 32)

If we look at the matter from the Ante-Chamber side, it is 2021.135 – 2046 or also 25 years. This is fittingly parallel to that King Chamber. The year 2021 is 500 years after the separation between the Luther Church and the Catholic Church.

7.4 Well of Life

Next thing to notice is the "Well of Life" and understanding the picture after over 26 years suddenly understood the complex picture of inside Pyramid passage system. While smiling to friends, saying this and adding PS and FP had a debate for 20 years about the Cubic, and AR correcting both in that way that these outstanding best specialists on the subject both misunderstood. Thinking that perhaps no one had really had this understanding (for few weeks) of well-shaft showing symbolic Jesus Christ open a new way for mankind, but see now that RM did first and P. Smyth confirming his understanding. This is 153 years after RM discovering's.

The Well-shaft, under the corner of Christ 26°18′9-7′. Below is Displacement factor meaning sin, and above is a rectification factor explaining the payment of Jesus Christ of that sin for humanity. Other pictures are showing the passages and their meaning and calculations.

AR Had the same understanding. Egypt text explaining Well of Life "The Crossing of the Pure Roads of Life or The Crossing of the Pure Water of Life[3]".

This needs further explaining (Figure 16-19), first as well-known, Jesus Christ was Crucified (Crossing) and Pure

Roads (or pure ways) and Pure Water of Life (as explaining the Well of Life, water has well known symbolic meaning truth or pure truth) and the last word (Life, the way is to Live though Jesus Christ) This is also an understanding of RM, it is correct mentioning that at the time R. Menzies had no information concerning that the Bible was fitting in any way to the GP story. Great work therefore and revelation of understanding concerning this matter, PS, and later AR, who adds the word: "obviously", showing this understanding. Now coming to the same conclusion as these gentlemen did before me, but it took me 26 years to find out.

Anyway, the comment of AR that this is "obviously correct" that the "Well of Life" symbolized the breakthrough of Jesus Christ through his crucifixion. After this is explained thoroughly this comment of AR is to the view: acceptable, after it has been explained, this is not even accepted today, without any sensible reason likely because of wrong End Time calculations of Pyramid specialists.

I simply see a picture that is not explained thoroughly. PS and Sir Flinders Petri had a debate for 20 years, concerning the cubic. As these were 2 best in the World in their field.

We see the Christ angle, and over it is rectification 286 number symbolized that Jesus Christ is paying for the sin of humanity and opening a new way through the Well-shaft from the descending passage which goes to the Pit.

"Well of Life" under passage is Displacement factor 286 explaining the sin of humanity, and Rectification factor 286 correcting and Jesus Christ is paying for the sin of humanity. It is correct adding here that Concentration factor or -35.76277 is in the Pit area and also corrected in the big step called Expansion factor or 35.76277 close to the King Chamber to the high way to holiness. is showing the Dead End, or if people do not accept the "saviour" a word used for Jesus Christ then people go down to the Pit. This means that people that reject God, simply are deleted called "The second death". There is no pain in that, AR explains.

The result of the specialist's working with AR, explaining that we have time until 2994 to accept or reject Jesus Christ. How so? This is done through reincarnation. This is AR and his cooperative people result. People that even have not heard the name of Jesus Christ, get another chance.

7.5   Granite Leaf or "The Boss"

Notebook concerning following 10 February 2018 and informed FB, SO, HKÓ and sr. Bjarni Rögnvaldsson about connection to or possible 3rd Temple matter, that is something drastic would likely happen 27 April -27 August 2019 and because of 1 inch minus discovered by PS something drastic would happen in 2018, and there are witness to the findings, but the plan was to go to Egypt and measure the GP and confirm calculations. These 2 points were seen by me, not marked in the picture, but looking at Anti-Chamber

Figure 48: The Temple coin-frontside.

picture (Figure 32) signed by DD (1921) and updated later, made (and confirmed) in Egypt in Ante-Chamber.

While looking better that we go under the Granite leaf in the Ante-Chamber. PS (died 1900) is mentioning long before DD drawing (1921) that under "The Boss" (Granite leaf) is shortening of 1 inch = 1 year. The conclusion that perhaps something would happen this year 27 April to 27 August (2018) that would be drastic concerning prophecy.

Not knowing that 14 of May the American Embassy would be set up in Jerusalem, on the 70th year of Israel reborn since 1948. One can say that America is the World Boss and perhaps they have also secretly setup 3rd Temple also at the same time. The idea was to go to Egypt and recalculate the matter, and also go to Israel to look at the American Embassy in Jerusalem. (Figure 48 and 49) showing silver coin made to enter the embassy, with a picture of Donald Trump president of America, with Cyrus in the front, and on the backside is a picture of the Third temple.

These calculations are following, the average year between 2021-2017 which is 2019 and because of PS finding of one year shortening it was 2018. Doing measurement's saying 27 April – 27 August 2018, something drastic would happen. It did the setup of America embassy in Jerusalem, and perhaps Third Temple it is interesting to look at the silver coin made concerning this

Figure 49: The Temple coin-backside..

matter and America is admitting that Jerusalem is the capital of Israel, not any more Tel Aviv. This is discovered by insight. The understanding was correct. One of 3 future calculations made. Appendix B.

7.6   The Queen Chamber the correcting room

Likely it was Charles Russell who explained first the Queen's Chamber as the correcting room. He suggested that the King Chamber is "high way to holiness" explaining that GP is showing the way of mankind prophesized the fifth Kingdom. GP is also showing the sin of humanity. This essay is correcting errors, but also confirming various results. Appendix B.

Anyway, while looking at this matter the number 1521 (Figure 32) is also in the Queen's Chamber, this has not been noticed by FB. This number is in third places, in the descent passage, in a narrow way, and in the Queen's Chamber. What is interesting here is that those who have been working together here concerning this work are mostly in the Church of Luther was fully separated in the year 1521. Part takers in that church are taking part in correcting this matter. Using probabilities, you have quite a low number of Luther church members compared to all people on the planet.

Then even more interesting is that the death year of Sir Isaac Newton is there also or 1727, which is taking part in the correcting this matter also.

So many information's in the GP that can be said that this is just lucky?

"Bible in stone" is likely to explain this matter. That the only person knowing about future is God and is therefore also designer of the GP this is the result of Pyramidology specialists.

Benjamin tribe: It was the hypothesis of AR that Icelanders were the Benjamin tribe. This is fitting Sigfús Elíasson books also, out that in that book mentioning, that the GP is "Pyramid of truth" therefore come up with End time calculation.

SB has traced lineage from Abraham - Isaac - Jacob - Benjamin - to Icelanders. Icelanders seem to be descendant of Abraham. Icelanders are Benjamin tribe, the rest seems to be lost, except Icelanders abroad, Faro Islanders and portion of Canada. Appendix A.

Tracing ancestry by (SB) is first introduced in this Ph. D. essay.

7.7   Benjamin tribe and the Bible

Finding out concerning the Benjamin tribe and the Bible. Romans 11) Reference[15, 25-28)

The Remnant of Israel

11.1 *I ask then: Did God reject his people? By no means! I am an Israelite myself, a descendant of Abraham, from the tribe of Benjamin.*

2 *God did not reject his people, whom he foreknew. Don't you know what Scripture says in the passage about Elijah—how he appealed to God against Israel?*

3 *"Lord, they have killed your prophets and torn down your altars; I am the only one left, and they are trying to kill me"?*

4 *And what was God's answer to him? "I have reserved for myself seven thousand who have not bowed the knee to Baal."*

5 *So too, at the present time there is a remnant chosen by grace.*

6 *And if by grace, then it cannot be based on works; if it were, grace would no longer be grace.*

7 *What then? What the people of Israel sought so earnestly they did not obtain. The elect among them did, but the others were hardened,*

8 *as it is written:*

God gave them a spirit of stupor, eyes that could not see and ears that could not hear, to this very day.

9 And David says: May their table become a snare and a trap, a stumbling block and a retribution for them?

10 May their eyes be darkened so they cannot see, and their backs be bent forever[15].

After this finding: FB had no doubts any more. He had been follower of Charles Russell, but after seeing this he changed his opinion concerning the Benjamin tribe.

This tribe seems to be the only tribe of Israel that was not rejected by God.

7.8   Probabilities of chance

Probabilities of a chance of information in the GP.[37]

As one can see the information's are so massive and the main conclusion is that the probabilities are less than $(1/10)^{52}$

This is done in the following way and not all information's are included.

Accuracy is every number given or (1/10) from 1,2,3,4,5,6,7,8,9,0.

3 years accuracy is at least 6 numbers in each year (in some places saying 8 numbers to take less chances) or in total (1/10)18. Figure 1.

Average way to the Sun or 92.996,1 miles or $(1/10)6$ (D.D) and Pi or 3.14159 $(1/10)6$, then speed of light 9 numbers or 299.792.458 $(1/10)9$. Pyramid angle 51°51'143'' $(1/10)7$ and Christ Corner or 26°18'9-7'' $(1/10)6$ in total fewer possibilities of all information's in the GP than $(1/10)52$. Original guess was less than. $(1/10)^{100}$ To do an accurate calculation computer program must be used.

Everyone has bad ideas sometimes; we must say when we are wrong. In this matter, in the view, this result is so obviously correct that it is over the only fragment of doubt. For those who still count on this very tiny fragment, is it sensible to trust on a fragment far smaller than 1%, changed to $(1/10)^{52}$ or Euro win 8 times in a row is more likely than the information in the GP. Part takers 6 million people and just 1 win each time.

"He has weighed the world in the balance, and has measured the times with a measure, and carefully counted the hours, and he will not move or disturb them until the prescribed measure is reached. "(II Esdras 4: 36-37)[15].

Afterwards, the specialists in the GP have not thought of the possibilities of all the information in the GP. We live nowadays in another time and therefore these calculations were made. One can say that there is a very small fragment of possibilities that the GP has its information by chance. Using the methods invented by Sir Isaac Newton in Calculus. You can have a number infinitely small, but not zero. These methods are used in various fields, and Newton started to think of an apple falling, and what is the speed of the apple. Some specialists say this story of Newton is made up. Strangely in the Bible, it was also misuse of an apple that brought sin into the world. The tree of good and evil was forbidden to eat an apple.

Religion is based on trust and belief. That trust must be looked at from a critical point of view. Atheism is one of religion's some people trust in.

The longest line on landmass goes through the GP. How likely by chance? Pi or 3.14 is shown in the GP. The base of the GP is fitting the days of the year or 365.242. Speed of light is shown in the GP under another measurement. Egypt people used 365 days in a year, so this was from Enoch's descendant controlled the building. GP is in harmony with the constellations Orion, Sirius, dragon star, Pleiades or 7 stars.

JT views that the GP is built under the "squaring of a circle" mathematical law. Only under the angle 51°51'143". How likely by chance?

DD and FB and AR result concerning the GP are agreed on GP (GP) has the measurement 1 inch = 1 year (Pyramid inch is 1,001 British inch). As the author has shown in Israel- Britain book British inch fits the inch in the GP and also the USA inch. One can say that the measurement fits the Icelandic one too that is the old measurement. The conclusion of the author was that the GP had a special meaning for those 2 countries. These nations have kept a guard around Freedom which is also fitting into the will of God. Main publishing of the Bible is made by people in the UK and USA.

GP shows 3 different lengths of the year shown in the book with 6 numbers accuracy.

- 365,242 Solar year

- 365,256 Sidereal year

- 365,259 Anomalistic year.

"The ancient Arabic Akbar Ezzeman MS., tells us something of its purpose. It states that the Pyramid contains "The wisdom and acquirement in the different arts and sciences... the sciences of arithmetic and geometry, that they might remain as records for the benefit of those who could afterward comprehend them... the positions of the stars and their cycles; together with the history and chronicle of time past and of that which is to come[1]"".

Four blood moons could be seen in 2014-5 in the feast days of the Jews. In 2018 and 2019.

3 years accuracy is at least 6 numbers in each year (in some places saying 8 numbers to take less chances) or in total (1/10)18. Figure 1.

23 September 2017 we saw Revelation 12[21].

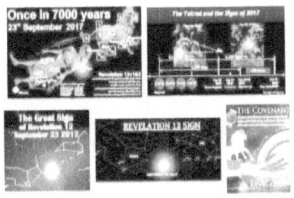

Figure 50: Revelation 12.

7.9   Big Phi

(N) Big phi (my finding ET) or 222222/777 = 286 small Pi is about 22/7.

It took me more than a year (1 ½- 2 years) to find out this Mathematical formula which is a basic understanding that 286 is not Error. This number is too big, and this was the key to this understanding. It was Dr Charles Lagrange who gave this number the name Displacement factor and called it the error of the GP. It was not until DD said that this is in too many places in GP to be an error. Discovered in the year 2018. In early books of AR, he uses the word error but later finds out that DD correctly discovered that this is in too many places.

7 x 286 =2002 was basic for End Time calculations. 286 + 2 = 288 and 288x7 =2016 +/- 2 years. (online 2005 put in big phi hugi.is, Elo8)

To find out a new formula, can be a lifetime work of a person, and taught in schools in a few minutes, lacking all depth of research. This is why returned in Mathematical department. This is mostly Mathematical and Statistic is just part of Mathematics.

10

Historical views discussions concerning this essay

PS (PS) was a professor in astronomy and a scientist. He went to Egypt and made exact research. In his conclusion, he confirmed the idea of RM and JT. PS (PS) also discovered that inch = 1 year and he wrote a book "Advent of Pyramidology" 33 Inch centre is showing the years of Jesus Christ. PS found also scored lines 2141 BC in the Broadway passage which leads to destruction that is shown by the Pit and this fits the year 2623 B.C. Shining into that passage is the Devil star. Astronomy was PS special field and was useful concerning looking up to the sky and look if there are connections between Earth and the sky. The Moon and the Sun. The designer of the GP knew more about the Universe than anyone else on the planet. PS was not a specialist in the Bible, but he was an outstanding specialist concerning Astronomy. PS confirmed that the GP is in fact "Bible in stone"

This interesting result of PS discovered by many years research showed people that not only was possible to look at the past in the GP, fitting to various historical events but also it was possible to look at happenings in the future (by using 1 Inch = 1 year) That is if the GP was correctly understood.

Charles Russell was in contact with PS and had been doing research concerning the Bible and he also had similar ideas concerning the passage system, he was happy finding out that this matter is in fact confirming the Bible. Charles Russell in his books about the Bible mentioned the GP work, and also very important is the Divine Plan of the ages.

Dr Flinders Petrie went to Egypt and did measurements of the GP and he is sometimes called "the father of archaeology". He was very exact in his measurements and Dr Flinders Petrie was an atheist. He also had a different opinion about the cubic size. There came up disagreement between PS and Dr Flinders Petrie and the main conclusion was that measurement's confirmed each other. Because of this debate, formal result in the matter was delayed. Dr Flinders Petrie was knighted 1923.

Many years before 1914 people had opinions that drastic time was ahead, and it would take about 4.4 years. Some thought 1913, others thought 1914-1918 correctly. As AR mentioning in his book a person from Canada had discovered about the length of the WW1 would be 4.4 years, but as he had calculated wrong years then his findings did not quite succeed because of wrong time frame as his finding deserves.

Dr John Edgar and Morton Edgar did measurements concerning the Great P. They discovered that the air shafts of the Queen's Chamber, fits the constellation of Sirius and they agreed with PS and JT and RM the GP was "Bible in stone" They also discovered Christ angle in the GP, that is the angle from the GP showing where Jesus Christ was born. This calculation was later confirmed by Captain John Mace ague and William Orr Warden (in the introduction[8])

The GP also fits into:

Isaiah 19:19-20 (Bible)

"In that day there be an altar to the Lord in the midst of the land of Egypt, and a pillar to the Lord at its border" AR's father James Rutherford, discovered that when this verse is put to numbers the total number is 5449 that fits the high of the GP. Work of James Rutherford older[15].

The shape of the Nile delta fits to circle and in the centre is the GP. The person who discovered about the shape and then discovered that the GP is in the centre. "This is the most importance building in the World"

Jesus Christ opens (symbolic) a new way for the human race by the way of Life with his sacrifice and paying the price of the sin of Adam and Eve in the crucifixion, with his blood. This is the Bible story and fit's GP story, and it is astonishing symbolism, and in fact obvious when looked at thoroughly.

Behind the Bible (God's word) is mathematics as shown by Dr Ivan Panin, who went over the Bible to his result is divinely inspired, for example in various forms of multiple of the number 7. As the Jews know the Cabala wisdom and the Jews are no fools. Why do they not look at the New Testament from that angle which is also inspired as the Old Testament?

Now talking to Avi Lipkin (Jew, was in Iceland 10 May 2018) He knew that behind the Old Testament were numbers, because of Cabala, but it was news to him that it is also in the New Testament.

According to AR Jesus Christ was born 29 September 2 B.C. (Saint Michael day or Feast of Trumpets). And that fits the GP result. Christmas is celebrated 9 months earlier when Mary got vision concerning Jesus Christ.

HA specialized in the Bible prophecy and he agreed with JT and PS but corrected P. Smyth concerning the measurements should take notice of the crucifixion, and not the birthday of Jesus Christ only. He and DD structural engineer started to cooperate. It is, to say the least, strange that they could work together because DD was an atheist, willing to disprove PS result, and DD more agreeing with Dr Flinders Petrie. Later DD discovered that his opinions were completely wrong, and the info in the GP was impossible, this was sort of "mission impossible". HA died in 1918. They published "The GP It's divine message"

J. Clemishaw corrected HA work. The basic result of these gentlemen was: GP is "Bible in stone" confirmed.

The book is very complicated and careful reading is needed. Because of DD background, he, later on, went over his calculations, some suggestion of others. These recalculations were useful, as through his work End Time as later shown.

"There are two matters to which Mr. Davidson has asked the writer (J. Clemishaw) to make special allusion. He wishes to lay stress upon the extreme reluctance he experienced in adopting a new horizontal scale for the Ante-Chamber, King Chamber, and connecting passage. This horizontal scale was suggested by Mr. William Reeve of Toronto in 1904 and adopted by Rev. commander Roberts, who also discovered that the perpendicular from the still of the entrance Doorway of the first low passage leading to the Anti-Chamber indicated the date of the beginning of the Great War. (WW1) Mr. Davidson was hostile to the adoption of this scale, investigating the matter found that if the unit were taken as a month of thirty days (instead of one-twelfth of the solar year, as adopted by Commander Roberts) the theory actually agreed with the geometrical indications of the Ante-Chamber, and also defined the dates of the beginning and end of the Great War. The other matter has reference to the Pyramid's representation of Natural Law and of the ancient Law of Relativity[8].

AR is in fact the last great scientific person in a full-time job doing research concerning the GP. The father of AR was also interested in the GP and discovered numerical 5449 which is also the high of The GP in inch in the Bible in Isaiah 19 19-20. AR discovered many details about the GP proving mathematical views of various matters concerning the building. He discovered 1925 the 5ᵗʰ socket and lines showing Iceland between the lines. Published Israel Britain and "Iceland light to the nations" In this small booklet, AR shows that Iceland is Benjamin tribe strongest. AR agreed with JT and PS and RM. Pyramidology 1-4 is his great work, and engineer James Rutherford carried on AR work. FB took over and published Pyramid Quarterly, holding the torch. FB is a big reader and specialized in the Bible but knew also about AR's work and his son engineer James Rutherford.

"In the design of the GP itself over 2000 years before earlier it is revel to a very high degree of precision[3)]".

Pi in the GP was more correct than 2000 years later was known. The probability of such information is also almost zero.

11

SHOWING VARIOUS YEARS

GP is showing various years in the history of man.

2623 BC year fitting the passage of the GP to a constellation. AR thinks this is the planning year and 10 years later or 2613 the work started, and it ended in 2589 or taking 24-34 years, perhaps when the GP was built. AR is at least much closer to the building years than today specialists. AR takes simply the lines to the stars and according to his info scored lines 2141 first discovered by PS, but he had measured 2140 +/- 2 years is well within acceptable scale. These calculations are confirmed by historians as discovered.

2141 BC scored line in the GP discovered by PS. These findings were not confirmed until 1938 by specialists. GP said beforehand 4500 years before. According to AR finding.

- 1453 BC Exodus.

- 33 AC life of Jesus Christ.

- 1440 AC printing starts.

- 1521 AC Luther establish his church.

- 1844 Great step (also in Ante-Chamber) to End Time.

- 1914 AC WW1.

- 1939- 1945 WW II.

- 1941 Iceland independence preparation said by AR in his book 1937.

- 1994 Hypothesis of End Time, ET.

- 2002 End Time starts, ET.

- 2006/7 drastic time Bank fall (though they could delay until 2008 (ET).

- 2015 23 September Sir Isaac Newton calculated concerning the Catholic Church gets more power.

- 2016/7 End Time calculations end (just prophecy to fulfil ET) (and Newton found 2016 also).

- 2000-2050/60 End Time calculations Sir Isaac Newton (later 2016-2060)

- 2018- 2051 hypothesis by (ET) mostly work seen in Ante Chamber (D.D update) and also "the Boss" findings by ET.

- 4044 BC-2045 DD mentions this time hypothesis.

- 2994 after peacetime period (AR finding).

- ET discussion's about prophecy.

- Then evil-doers go to the Pit.

Newton was afraid of saying about End Time (2060) because people might stop believing in God if the calculations are wrong. As somewhere is mentioned End Time means the end of the period in world history. According to specialists Bible will be taken more seriously, and afterward, a more peaceful time will come.

Reading about Sir Winston Churchill the great British minister and standing against Hitler. It is in fact amazing that Germany could not break down Britain, and it is known that the great politician was an inspiration both to others and his strong belief was based on this Pyramid knowledge, that Hitler would lose. From a strategic point of view, even working with Stalin to crush Hitler was incredibly profound and then Hitler lost against the Russian winter.

Newton came to the conclusion that behind the Bible was mathematics and he was trying to find a code about it. In Bible code books stand that because of complexity "computer" was needed to solve this matter.

It was the famous economist John Maynard Keynes who bought 1936 the papers of Newton then later on published online seen at the end of the essay. References [9]

Newton discovered this through deep investigation of the Bible and he wrote a book about Daniel and Revelation of the Bible which had to do with the End time. Newton went into translation of the original Bible papers (this is very complex matter) End time period of Newton was 2000-2050/60 depending on publishing of books.

Newton calculated 23 September 2015, correctly.

23 September 2017 Revelation 12 in the sky. It is mostly misunderstood.

23 September 1845 is 172 years before this advent. Called the Autumnal Equinox, going through the Queen Chamber and the great step.

"The intersection of passages and chambers by the plane of the Pyramid's east-west axis is always at point marking the Autumnal Equinox, whatever the year and whatever the chronological scale involved[1].

Looking at the Pyramid books, they were talking about the error 286 or sometimes called displacement factor. In the Grand gallery is rectification factor or same number correcting. This is done with the sacrifice of Jesus Christ crucifixion (correcting people's error of sin).

Correct understanding of 286 is that there is not (big) error in the building itself, but this is error concerning human kind and following the creator.

Here is correct mentioning that over "well of life" 286 or Rectification factor is correcting Displacement factor -286 through Christ[2].

Contraction Factor (-35.76277) is also following displacement factor of 286 / -286 and expansion Factor is (+35.76277)[2].

In the year 1994 coming up with hypothesis that End Time was 2002-2016/7 286 is called symbolic of sin and therefore 7 x 286 = 2002 and correction 286 + 2 = 288 (or some say the note D) explained or 2016. FB was informed about these hypotheses. In one book from AR (later discovered) he is mentioning 1994-2994 time period. 7 x 286 = 2002 and also 8 x 286 = 2288[2].

Not knowing what would happen, but then while time was ticking, discoveries like 10 years of shortening would lead to 2002-2006/7 something drastic would happen. It was not until 2008 the Bank fall of 3/10 biggest bank falls in the world, were here in Iceland. Then later discovering that the Bank fall was in the year 2006 but was delayed for 2 years. Sometimes 2 years of delay comes in this matter.

Then the latter year 2016, found in Ante Chamber 2016, 257 that is 3-4 of April that year. Pyramid friends where informed about that day. Here is info of importance. The years 2002-2016, and FB was mentioning this is not so exact what Fred said that only calculated years not up to day. Then in the beginning of 2016 this is in Anti Chamber picture 2016,257 fitting into calculations and then discovered that this was 3-4 of April. Panama papers were the main news that date.

DD seems to have updated his calculations from his book with HA and not mentioning anything concerning that matter further. Ante Chamber is showing from year 1844 -2051 and seen 19 points and have calculated 2 more. ("The Boss calculations) That is hypothesis is confirmed.

What happened 2016? That is on 3-4 April 2016. First, this is the Crucifixion date 3-4 April in the year 33 of Jesus Christ. Panama papers were published formally 3-4 April 2016 see end of essay.

This seems to me showing Icelanders:

The politicians and part takers in corruption is judged, globally.

First bank fall 2006 and then Panama papers 2016 showing that many Icelanders have money in hidden source (around 600 persons) Putting in end notes from the Internet video, it does not say anything about the ex-pr. Minister, perhaps he was not guilty. Some more guilty has got away with that until now (2018) globally bank crooks, got away with crime, the situation is terrifying. In Iceland some of them were put in jail.

Icelanders are therefore specially taken out and judged by their corruption. Judgment comes first on Israel and as Icelanders are Benjamin and Benjamin part of Israel then this fit.

A lot of prophecy has happened in this 2002-2016/7 for example 4 blood moons on Jewish dates and then symbolic wolf in the beginning of year 2018 and then 31 January huge blue moon. Now in January it starts 21 January also 2019 wolf moon is the first of blood moon and bigger (closer to Earth).

Looking back at Newton's work. It seems that he changed his mind in later years to 2016-2060.

11 of February and 1 of November, the GP was shining when the white stones were there, but Arabs took them away from the GP. Perhaps on the top was golden symbolized cornerstone

Shows detailed information's about the Icelanders fitting into both Bible and the Pyramid mathematical result. All the true apostles were of Benjamin tribe, and the first church was established in Britain in Glastonbury and both uncle of Jesus went, Joseph of Arimathea who was a tin seller who took down the body of Christ. Apostle Paul went to UK and the Paul church is setup in Ludgate hill.

Looking at Anti Chamber backwards, dates are back to 1844, and 1879, 1941, 1992, 2006, 2016

Ante-Chamber: Is sometimes called architect room. It is explained to be 5 Royal cubic long, and first low passage is 2 Royal cubic, and 103.03296 Inch. DD calculated the Ante-Chamber and his picture since 1921 is later updated it is not known if he did the update himself thought it is likely as his name is under the picture and likely did that. These are just measurement and no explanation included.

Ante-Chamber has also Granite leaf sometimes called "The Boss". PS explanation is very interestingly accurately made. Fred Bins said the calculations are rather primitive just years not date. This can be understood by reading AR books. King Chamber has 103 and 230 numbers multiplied with various numbers.

12

THE DIVINE PLAN

The Divine plan of the ages portrayed in Pyramid symbols (p. 1349) References [4]. In the time of the End prophecies will be fulfilled and fitting into plan. Scientific result and Divine inspiration will be harmonious in Religion. (Habakkuk 2:2) "GP seems to be Bible in stone".

"The Battle of the Great Day of God will have a terrible beginning but a glorious ending[1]".

On page 1337 References [4] that William Reeve, did calculations concerning Ante Chamber and had hypothesis concerning WW1 52 Inch would mean the WW1 war would take 4 years and 4 months, his calculation concerning WW1 was 1910-15 and therefore he did not get the attention as his theory did deserve.

PS "In a letter one of the most eminent British astronomical authorities has assured the author: Not only that 2141 BC is the correct year for transit of the draconic in this connection, but that even the year immediately preceding and that immediately following, 2142 BC and 2140 BC respectively, are both not only outside, but "well outside the limits of error permissible[1]".

It is of great importance that what will happen, the future will tell clearly, and the symbolism of the GP is not an easy task to understand.

Another one Professor, F. Petrie, who was very exact in his calculations and the father of Egyptology, and archaeology research methods. Well known is the pottery research of Flinders Petrie who discovered timing with that method of history.

Dr John Edgar and his brother Morton Edgar and his brother calculated the passage system, and many others have been doing deep research of the GP.

From 1900 mostly two groups have been doing serious research concerning The GP from this angle that is fittingly into Bible views.

The other two were HA who was doing special research concerning the Bible and prophecy. Strangely he was working with an atheist DD who was a structural engineer. After HA death 1918, DD carried on the work. Writing both of them for their main book[8].

DD discovered that the information in the GP were impossible except through Divine control of the matter. He therefore *changed his mind.*

Likely DD's recalculations of the Ante-Chamber, not saying anything about his view, were useful for me in this matter. DD died 1956.

The last two were AR who did a huge work concerning this matter and James Rutherford who took over his father work for some years.

These two groups came to the conclusion that yes End time was from 1914 to 1953 20 of August and doing also calculations into the future. As nothing special expected happened around 1953 and later put later to 1979 most lost interest in their theories and views and thought that this matter had been misunderstood in a way at least.

AR calculated $7 \times 286.1 = 2002.7$. "How appropriate that the rejection of the great "Top-Stone", the sinless Lamb of God, Jesus, by sinful humanity should be symbolized by a number that is sevenfold the symbol of sin!" Then AR adds that 8×286.1 is 2288.8[2]) or rectification factor. How close he was to find out about End Time period. This was not noticed until long time after the calculations. Though AR had tried to write his books for "the man in the street", but also fulfilling scientific demands. These information's are so massive, and to get the point always one must both read carefully and many times, at least in this case.

AR was interested in Iceland, as in the year 1925 he found the 5th Socket and Iceland was between the lines who went also through UK and Iona in Scotland and was well known for Christian mission. Icelanders are fitting into Benjamin tribe theories.

Coming up with hypothesis of the End Time 2002-2016-17 and then 10 years shortage also would say 2002-2006-7 something drastic would happen, so the hypothesis was ready 1994 informed to FB.

End Time: What is that? That is a time when the Book of Daniel and Revelation last book of the Bible go to fulfilment. There will be short difficult time, but then afterwards better times are coming, and prophecies will be fulfilled. Already from 1994 for example 4 blood moons, on Jewish dates. Financial crisis 2006 and also 2008 and 2016 global scandals have been seen. Ethics have fallen and corruption have been a problem. Bethlehem star seems to be returning, which is in fact 2 stars (Venus and Jupiter)

Looking at the books of Sigfús Elíasson, saying: "the Pyramid of truth".

Specialists were on the other hand talking about Displacement factor 286 as error which according to their theories was mistake in various places, later on 286 were found in many places in the GP. DD, he disagreed about this matter 286 were in too many places, but result was that 286 meaning was sin. AR agreed later on.

Reading about Sir Isaac Newton's work he did in the book of Daniel and Revelation his conclusion was that the Catholic Church did not have the authority it claims.

Calculation for 2002-2006/7 showed the economical crash, which is today admitted happened in the year 2006, they could delay this to the year 2008, and 3 of 10 biggest banks in the world, collapsed here in Iceland.

In the beginning of the year 2016, the discovering about the Ante Chamber that is 0,257 which is 3-4 of April. The day came up, yes this was the day Panama papers were brought up to the public. This is also the Crucifixion date. Icelandic prime minister had to resign.

Sometime there seems to be 2 years delay, and also of great importance concerning the types in the Bible as FB has shown me clearly. This means some story in the Bible not obviously understood has some meaning later on for some other purpose. This is one of very deep learning from FB. Scientists make various judgments without having a clue about the GP. GP is showing very clearly about the size of the Earth, length to the Sun and the Moon and various impossible information's about the planet Earth. Only one knowing about all this is God himself, this is the result concerning this matter. This is also the result of Pyramidology specialists.

Is this matter fitting the Bible?

Yes, very clearly showing that Isaiah 19.19-20 seems to be fitting to the GP.

Dr Ivan Panin (Doctor in Mathematics from Harvard, born in Russia) had made a research concerning the Bible. Behind the text is numbers and if the Jews would have checked this, they would have seen also that it is not only the Old Testament that is inspired by God but also the New Testament.

As AR shows in his book and uses text from the New Testament, which is variously multiplied by number 7.

The view of the year 1881 which is the measurement of the Grand gallery would be fitting in some ways to PS life. PS was himself trying to find out about matters, and even expecting something drastic happening (1881) in his life, which did not happen. Others discovered about 1914 many years before it happened, time, as usual, tells about such matters. Concerning dating 1953 this is the last Queen before End Time. King's Chamber symbolized Holy of Holy or the Tabernacle. AR mentioning that if you add years of Jesus Christ (33 years) to 1881 then we have 1914.

Concerning the time frame 1992-2018. Was this understood correctly? Yes. We are in the End Time and as FB says, we will see further progress, but the main thing is that this time frame is correctly understood. Newton was even more correct than me. Calculating years while Newton days (23 September 2015) this 3-4 April 2016 was found in Ante Chamber.

It seems that the basic result is that the hypothesis of the lost tribes is correct. "Why has the land Denmark this name"? Only 1 of 40 had a clue. That person was working in a pharmacy and said, "Has this not something to do with the Vikings?"

That was a huge surprise to me how little the Dan-tribe people new about their own history. In the beginning of the year (2018) there was unusually huge Moon. Symbolized with the wolf according to information's wolf is Benjamin symbolized (also 2019).

The first church was established in UK Glastonbury, it was Joseph from Arimathea, rich uncle of Jesus Christ, who got the body of Christ to burry in his tomb. It is well known that Saint Paul original Church was established by the apostle of Christ. Joseph from Arimathea was according to SB, the forefather of most Icelanders born here.

The intention of AR was to add one more book. Basic result in the matter is already there in 1-4 books on Pyramidology, though some errors are still there. The intention is to write a book after this essay is published. In the 1-4 books, information's are there about what ideas and discussions AR intended to discuss in his last 5 book. Concerning calculations then his intention was to add one more accuracy number to the scientific calculation's. Appendix C.

AR intention was to publish more about the lost tribes, the person financing the work disagreed. It is not wise to base research work in religion matters against the opinions of financial supporters. His calculation of 1953 and 1979 has time disproved and therefore people thought that the whole matter was wrong

AR agreed with PS that the British are holding the crown, and due to their good/ bad times they have not changed basic rules concerning the Kingdom. The measurement of

Egypt inch is very similar to the British and American Inch. The British have also Stone of Destiny and they seem to be correct concerning these matters according to the findings.

"Where the spirit of the Lord is there is liberty" 11 Corinthians (3:17)[15].

Today Queen of England has shown in history that they are of Israel origin. This is also clear when symbolized in Union Jack colours are of great importance here as in the flag of America. One outstanding hymn is Jerusalem Glastonbury. Queen is sitting on the Throne of God. Such information's are detailed on the Internet or in books on the subject.

It was Leif Eriksson (Icelandic) that first of Europeans to find America (Am Erica) in the year 999 (or 1000). A long time before Christopher Columbus. One of the most outstanding paper in the view is the Declaration of Independence while America was established in 1776. The bald eagle symbolized America, and fittingly.

This time frame 1992-2018 is just part of this theory. It is possible to see other time frame fitting into the GP, look up in Ante-Chamber for future points.

The result is also that the Ephraim tribe is mostly in the UK. Stone of Destiny is in Scotland castle of Edinburgh.

The Queen and King are anointed (that is to symbolize have the Holy Spirit) When Queen or King is insertion, they go throughout detailed rituals and with oil symbolized the Holy Spirit.

The second coming of Christ is explained: (and not what most churches are teaching)

"The mystery which hath been hid from ages and from generations but now known what is the riches of the glory of this mystery...which is Christ in you, the hope of glory" Edh, 5: 25-32, I Cor. 15:51-52, I Thess. 4:15-17, Col. 1:26-27[1].

13

286

In early books about this displacement, a factor is talked about matters like this is an error in the GP, that is an error by the author. Because of the complexity of the matter, it is not until now in the year 2018 was realized that these Pyramidology specialists changed their mind. DD structural engineer, and in the book "*The Great Pyramid its Divine message*" DD had in fact confirmed[8] that the 286 = Displacement factor, was in too many places in the GP to be not in line with the plan of the author

Later AR came to an agreement with DD though he is not mentioning that especially, DD result was that 286 were in too many places in the GP to be there just by chance.

DD changed often his opinions on matters like scientific thinking people do. In this final printing, his opinion is that 286 is in fact confirming the purpose of Jesus Christ in many ways. 286 is a symbol of sin, and Jesus Christ paid depts. of the sin of humanity with his blood sacrifice on the cross.

Lost tribe hypothesis[4]: Abraham, of course, was neither an Israelite nor a Jew. He was a Hebrew, as stated in Genesis 14:13. Israelites into existence when God changed the name of Abraham's grandson Jacob to Israel, whereupon Israel's (Jacob's) family and descendants constituted the Israelites. This was in the spring of 1833 B.C. It was not till after the division of the Kingdom into the two kingdoms, Israel and Judah, comprising the exclusively to the inhabitants of the southern Kingdom, Israel, and Judah in 938 B.C. That the term Jews came into existence and it was applied exclusively to the inhabitants of the southern Kingdom, Judah, comprising the Tribes of Judah and Benjamin. The term "Jew" is simply a contraction of "Judah" and was never applied to the people of the Northern Kingdom, Israel, neither in the Bible nor in secular writings. The first time that the name "Jews" occurs in the Bible is in II Kings 16:6 in the reign of Ahaz, king of Judah. At that time whilst all Jews were Israelites, all Israelites were not Jews. Just as today, all Scots are British, but all British people are not Scots. The southern

Kingdom was called Judah by virtue of its leading tribe being Judah the northern Kingdom was often called Ephraim because the leading[15] tribe in it was Ephraim (as, for example, in Isaiah 7:8).

Eventually the king of Israel (Ephraim) was deposed and the people of the northern kingdom taken captive to Assyria in the latter part of the 8th century and first half of the 7th century B.C., whilst the king of Judah was deposed and the Jews taken captive to Babylon in the early part of the 6th century B.C. It is not within the scope of this footnote to carry the details of this subject further, but it is a revealing and rewarding study to trace the House of Israel and the House of Judah through the ages of history[4].

14

LOST-TRIBES THEORY

AR's result in the Iceland "Great inheritance" was that Icelanders were the Benjamin tribe. Appendix A.

Dr Barði Guðmundsson: Icelanders came from Black see.

Back to the land of Israel, came both the Khazars Jews and the Ashkenazi Jews. This is agreed by the specialist Avi Lipkin (staying here in Iceland May 2018, for few days) He said also that the mountain where Moses got the law in mountain Sinai, which is in Arabia.

Dr Chuck M. who died 1 May 2018 and Avi Lipkin were friends. These are 2 of the best specialists lately, concerning religion's matter. Avi Lipkin understood the 3rd Temple, but most specialists do not have a clue.

Most Israelis live outside Israel. There is only part of Israelis in the land of Israel today and they are of Jewish descendant. There is great misunderstanding going on. Israel = 12 tribes. Most of these people went to Europe as can see in Table 4.

AR's result in the Iceland "Great inheritance" was that Icelanders were the Benjamin tribe. Appendix A.

Table 4: Tribes of Israel

Tribe	Land	Mother
Ruben	France	Lea
Simon	Silurian (Celts)	Lea
Judah	Scatterd, then in Israel	Lea
Zebulon	Netherlands	Lea
Issachar	Swiss/Finland Estonia	Lea
Dan	Denmark/Greece /Ireland	Bilah
Naphtali	Norway	Bilah
Gad	Sweden	Zilpah
Asher	Scots/Belgium/Luxemburg	Zilpah
Ephraim	Britain	Asenath
Manasseh	America	Asenath
Benjamin	Iceland/Faro Island/Canada	Rachel

All except Benjamin were born in Padan Aram (Kalisch).

First is to tell the story of the 12 brothers and Jacob who worked for 7 years for his beloved Rachel. After the wedding day, he discovered that he had married the sister of Rachel that is Lea who was the older sister. Then Jacob was offered to work 7 years for Rachel. Lea and Jacob had 6 sons. 2 sons had Rachel with Jacob, Joseph was the firstborn and when Benjamin. Rachel died giving birth to Benjamin. Jacob had 2 sons with Zilpah Lea handmade and 2 sons with Bilah Rachel handmade.

Joseph had dreams and his half- brothers were envious because of that. Later they cheated his brother Joseph and threw him down in a hole. Joseph was saved and then he was sold as a slave to Egypt. Because he could read and write he became assistant of Pharaoh's wife.

The High Priest's wife tried to seduce Joseph, who was not willing to do that. High Priest put Joseph in jail, and then later the Pharaoh had dreams that only Joseph could solve dreams are an explanation of God's spirit sometimes rejected by specialists.

The dream was following, he saw 7 fat cows and then later 7 thin cows. Joseph solved the dream puzzle and Pharaoh put him over Egypt.

The dream happened in reality, and Joseph sold the grain for other arias while there was famine in the land.

Then Joseph half -brothers visited Egypt to buy grain and Joseph planned to have revenge. He asked for his brother Benjamin who was not with the others, and later he set up a plan that the silver cup was in Benjamin sack. Benjamin got 5 times more than his brothers. Benjamin was not born when the half- brothers of Joseph cheated him. Benjamin took no part in that cheating.

Explanation of the tribes: from America (Manasseh) and Britain (Ephraim) in prophecy.

Lost tribes are showing in their flags and symbolized where they are from. Vital for an explanation concerning Benjamin tribe.

Reuben:

"Reuben, you are my firstborn, my might, and the beginning of my strength, the Excellency of dignity, and the Excellency of power: Unstable as water "(Gen. 49:3-4)[15].

Reuben dominates the Israelite element in France. France especially north-eastern France is heavily represented by the Ripuarian Franks, as bearers of the ensign of Reuben.

Simon, Levi, and Judah:

"Simon and Levi are brethren; instruments of cruelty are in their habitation. O my soul, me not you into their secret, unto their assembly, mine honour, be not you united: for in their anger they slew a man. Cursed be their anger, for it was fierce; and their wrath, for it was cruel: I will divide them between Jacob and scatter them in Israel" Are probably mostly in the land of Israel.

Judah only a fraction lives in the land of Israel, but also in the USA and UK and for example Hungary and Russia.

Zebulun:

"Zebulun shall dwell at the haven of the sea; and shall be for a haven of ships" (Gen. 49:13)[15].

"Yair Davidy states; "The Netherlands (Holland) features lions on its Coat of Arms, Lions symbolize all of Israel especially as united under the House of David from Judah whose special sign is a lion. The Dutch come mainly from Zebulon."

Holland is well known for sea transporting through Rotterdam etc.

Issachar:

"Issachar is a strong ass couching down between two burdens" Gen, (49:14) Issachar's coat of arms was donkey or ass carrying a heavy load[15)38).

Finland has been a battleground between Russia and Sweden, well known is how the Finns were fighting with the Russians on skiing, they also are famous for their stamina runners. Swiss is also Issachar land, and Estonia and Hungary.

Dan:

Dan shall Judge his people, as one of the tribes of Israel. Dan shall be a serpent by the way, an adder in the path, that bites the horse heels, so that his rider shall fall backward" Gen. 49:16-17[15). Dan is also in Greece, and perhaps in Ireland.

Dan River, Dan hill, Dan tribe is also in Greece.

Gad: "Gad, a troop shall overcome him: but he shall overcome at last" (Gen. 49:19)[15).

Gad is mostly in Sweden,

Asher:

"Out of Asher his bread shall be fat, and he shall yield royal dainties" (Gen. 49:20)[15].

There is a slight difference in various books concerning this matter, some say Belgium and Scotland are Asher strongest, Swiss and Luxemburg.

Famous is an answer from Scotland to the Pope, they knew they were one of the Lost tribes of Israel.

Most of the Pyramid specialists are from Scotland.

Naphtali

"Naphtali is a hind let loose; he gives goody words" (Gen. 49:21)[15].

This has been explained as Norway and blend to Sweden also.

Joseph

"Joseph is a fruitful bough, even a fruitful bough by a well; whose branches run over the wall: the archers have sorely grieved him, and shot at him, and hated him: but his bow abode in strength and arms of his hands were made strong by the hands of the mighty God of Jacob. "(Gen 49: 22-24)[15].

Father of Ephraim and Manasseh Joseph lived in the land of Egypt they were born in the next angle to Christ corner later shown.

Britain and America are obviously connecting to the GP their measurements shown by for example PS and also by AR.

Britain is having a correlation between the royal family of Britain and Judah's royal line, is found in Britain's emblem.

Stone of Destiny is showing clearly where God's throne is. While the Queen was starting 1953 the tradition is very firm and from Solomon's heritage.

"In scripture, Joseph the father of Ephraim and Manasseh is likened to a fruitful bough or branch, most likely an olive branch. Joseph is also described as being shot by archers. Later, Joseph's descendants are further described: "His glory is like the firstling of his bullock, and his horns are like the horns of unicorns; with them he shall push the people together to the ends of the earth, and they are the ten thousands of Ephraim, and they are the thousands of Manasseh" (Deut. 33:17)[15].

Manasseh is the tribe 13 is symbolized in various ways in America emblems.

Worldly play and today status:

Explanation of the World status, and for everyone to make his/her own judgment.

In this book of America and Britain in Prophecy (David C. Pack)[38)]

These nations have been loyal to God and spreading the message. There came then a new way of telling stories. Catholic Church has never been part of the Lost-tribes theory. It was strange to see suddenly that the Zara was brought into play likely by the Restored Church of God. Their meaning is that this church is the correct church. Future will show this very clearly before 2060, about this matter. To Catholic friends, we agree concerning Jesus Christ. Is there any church following the Bible fully? One of the Catholic friends has agreed that the original text is the basic. According to the findings, the church of Luther is the correcting church. Down the Descendant 1521 (Luther establishment church year) inch and years and up the Grand gallery 1522 (+/- 2 years) and then into the Queen chamber 1521 inch (and years) correcting the Catholic Church. Queen Chamber has also 1727 seems to symbolize a year of Sir Isaac Newton (within +/- 2 years) showing in the view that he is a part taker of this corrections Queen Chamber is called "Restitution of all things[1)]". According to AR and Pyramid specialists, Queen's Chamber is the correction aria in the GP.

America has the highest score as the most dangerous or 24% this is the result of a poll. That can partly be part of propaganda also because some Americans see themselves as the world's guardians.

Attack for example on Libya was a fraud and getting so-called Western nations into that this is often in cooperation with United Nations who have been misusing power. Fake news, misusing media, and in fact, and the Television is heavily misused for wrong information globally.

It would be good to inform about the situation today. This is more to open discussions than stating about the situation today.

Zara was added as one of the 13 tribes, then realizing that they had a plan, adding the Catholic Church into the lost tribes through Ireland. As can be seen, the 13 tribe is Manasseh tribe of America and can be seen symbolized in America flags, etc. since 1776.

Manasseh and Ephraim were one tribe Joseph (1/12), but was divided, the symbol of Manasseh is 13 and the Bald Eagle.

So what tribe were the light-bearers of the good news?

As all of the true Apostle were of Benjamin tribe, and they went to Britain to establish the first church.

America and Britain have been the most active nations spreading the Bible message.

Who was setup in Glastonbury? Joseph from Arimathea was the uncle of Jesus Christ, and Christianity blossomed from that source. Joseph of Arimathea is the forefather of most of all Icelanders born here. He was both of Judah tribe and Benjamin tribe. Likely the family of Jesus was having both Judah and Benjamin tribe blood.

It is correct mentioning concerning Jesus Christ that there is a difference in saying. Some say he is God, born in flesh. Some say he is the son of God. Newton says he is not God himself.

Just to give this explanation was needed for the last part of this booklet.

Namely Benjamin tribe:

AR who found with Edgar Morton the 5th Socket of the GP

AR discovered that Icelanders are in between the lines as can be seen here.

The same tribe is also in the Faro islands and also in Canada.

While Moses went through the see, the first ones trusting God were the Benjamin tribe. Then also seen that Benjamin had more children and therefore Benjamin tribe is bigger in the beginning.

Here is a 20 years' work of SB. Tracing back to Abraham (Benjamin tribe) and then later to Adam. Some people have different names, for example in Egypt and Israel language. Tracing back is specialist work. Just like some people are an accountant and some are salesmen and having different education.

Why do this?

AR has huge work concerning this matter. Knowing that the Jewish had written down their tracing back to Adam from the start of the creation of man. Hearing that there came up a misunderstanding or an error concerning that matter that has been corrected.

In the book of AR, his result is fitting into SB outstanding work. The GP is also confirming this hypothesis.

The following info from AR caused silence for hours and what FB said concerning the calculations that were in years 2002-2016/7 and not until 2016 finding of Ante – Chamber mentioning accurate date.

These calculations of AR are revolutionary:

Adam (created) 5407 to +2994 – 1 = 8400 years.

This is from a mathematical point of view play of the numbers 3 and 4.

That is 3+4 = 7 and 3 x 4 = 12 1200 x 3 = 3600 and 4 x 1200 = 4800 years.

Or put in another way 5406 ¼ + 2993 ¾ = 8400 years[1].

7 spiritual perfection and 12 as the perfection of government or governmental perfection.

6 means symbolized imperfection and 7 perfections.

AR has more to say about this matter 280 x 30 = 8400 and Jesus Christ was 30 years old offering himself as a sacrifice. 280 is a play of 4 x 7 x 10 = 280 and Rutherford conclusion is that the Queen's Chamber measures 10 sacred cubits fitting into the holiest.

The year1521 is symbolized Luther church correcting the church of Catholics. This number is not only in the Broadway and crossing 1440 printing starts, which is both parts of 288 (or 2x 144) and well-known number in the Bible 144.000, but also this is a new door for humanity to read the Bible against the Papacy will. Also discovered is that this number 1521 is also measurement to the Queen's Chamber and is, therefore, part of God's plan.

It is also of great importance, to tell the truth. AR went over the work of other Pyramid specialists. His calculations are more accurate than mine. While calculating these matters 2002-2016/7. What AR and his son James Rutherford calculated up to date.

In their book, correcting errors but confirming otherwise AR & son work[1]. Calculation of birth of Adam is there 3 October 5407 B.C.

Even calculation of day of the sin of Adam is calculated as 10 of Tishri or on the Day of Atonement. In our calendar, it is 19 of September. It is correct to mention that Jesus Christ paid for the sin of Adam while "baptism into death tracing is confirming Adam and therefore the work of these gentlemen. That is the Bible, Pyramid SB, and Pyramid specialists.

"on 10 Tishri in his 29 years."

To understand this one must learn about the Jewish days like Newton did in his days. This is in fact all a part of God's plan. The conclusion is therefore following.

AR his father James Rutherford who calculated Isaiah 19.19-20 is fitting the height of the GP also and his son use a huge time concerning tracing to Adam.

15

TRACING BY SIGURÐUR BJARNASON

We fulfil the laws, so every Icelander can go to the book of Icelanders, on the internet (islendingbok.is), and connect to Snorri Húnbogason.

Placing here the tracing: It goes from Adam & Eve to Snorri Húnbogason. Most Icelanders have him as a forefather.

Adam[39] & Eve, Seth, Enos, Cainan, Mahaleel, Jared, Enoch, Methusaleh, Lamech, Noah, Shem[15], Arphaxad, Cainan, Sala, Eber, Peleg, Ragau, Saruch, Nahor, Terah, Abraham, Isaac, Jacob, Benjamin, Huppim, Matri, Becher of Jemini, Aphiah, Bechorath, Zerror, Abiel, Kish, Saul, Jonathan, Meribbaal, Micah, Ahaz, Jehoaddah, Zimri, Moza, Azel, Azrikam, Ahio[40], Achaz, Ezechias, Manasses, Amon, Josias, Eliakim, Jehoiachin, Salathial, Zorobabel, Resa, Joanna, Judah, Joseph, Semel, Mattathias, Maath, Nagga, Esli, Naum, Amos, Mattathias, Joseph, Janna, Melchi, Levi, Matthat, Joseph of Arimathea, Anna, Penardim, Bran, Caractus, Cyllin, Coel, St. Lucius (Llewfer) Mawr, Cadwalla, Dr Frigg, Skjöldr[39], Friðleifr, Frið-Fróði, Herleifr,

Hávarr, Fróði, Vémun, Dr Ólof, Fróði hinn friðsami, Friðleifr, Fróði inn frækni, Hálfdan, Hróarr, Valdarr inn mildi, Haral, Dr gamli, Hálfdan snjalli, Ívarr víðfaðmi, Aud the Deep-Minded, Randvér, Sigurðr hringr, Ragnarr loðbrók, Sigurð snake in the eye, Ása (Álof)[41], Ingjaldur Helgason, Ólafur the white, Þorsteinn the red, Þorgerður, Höskuldur Dala-Kollsson, Ólafur „pá", Þorbjörg digra, Ingveldur Vermundardóttir, Þorgerður Yngvildardóttir, Yngveldur Hauksdóttir and her son Snorri Húnbogason.

Snorri Húnbogason is an ancestor of most Icelanders.

The GP shows human race crawling 6/7 because of Adam's sin and then 1/7 you can walk upright.

The triple cord is not so easy to break. (Ecclesiastes 4:12) The Bible.

AR shows the similarity between Icelanders and Benjamin tribe. He was likely the first one to discover this and from the GP. Icelanders have though kept more information concerning past that many other nations.

If we go back to the 12 sons. 6 (include Levi) had the same mother Lea, and all of these plus sons of the handmade of Lea and handmade of Rachel cheated Joseph, who later became a powerful person in Egypt. Joseph and Benjamin had the same mother Rachel. Jacob wanted to marry Rachel, but to his surprise, his sister was married to him first. Then he had to work another 7 years for Rachel. 7 has a symbolic meaning[1].

In the kingdom of Judah, there was Jews who rejected Jesus Christ as a messiah, and Benjamin tribe who accepted him and took no part in the Crucifixion.

Therefore, AR came to the conclusion that Benjamin tribe was the people of God. This can also be seen in the Bible. You judge from the fruits. Jews still reject Jesus Christ as Messiah, and the Benjamin tribe accepts Jesus Christ by a majority.

The books of AR Pyramidology 1-4 were not bought by me until just before the year 2000. Anti-Chamber picture was not found until at the beginning of 2016, and Newton's work (2000-2060) discovered by me 2017 perhaps put online before that time. New calculations are confirming Newton's calculations. 1994 was the hypothesis ready and then later calculated 2002- 2016/17 for me personally it is strange to say the least looking at Ante Chamber, when 1992, 1994 (hypothesis ready), 2002 (7 x 286), 2006 (10 years shortening or 2x 5 (grace)), 2016/7 (7 x 288) 2018 (+/- 2 years, 2021 and 2038 and 2047 and then finally 2051. See picture Ante-Chamber and look for yourself bigger size in this essay on the last page.

The Icelandic flag is in line with UK and USA flags, and the artistic painter Thorarinn B. Thorláksson was among others taking part in flag comity.

Icelandic symbols fit into Ezekiel's vision. As Jonas Gudmundsson says in Dagrenning 1 paper the symbols fit Ezekiel except the east part has a different angle. Lion changed to Serpent. Symbols are telling a story. Among

Icelanders well know is the spirit connected people as was well known among the tribes of Israel. This comes through visions or dreams in connection with the spirit[13].

Future tells if the visions are true. This was also a problem in the old days and for example, Jeremiah was thought to be a false prophet, it was not until later that it was agreed that he was a true prophet.

Jews reject New Testament and therefore also various information's in the New Testament has discussed with Avi Lipkin who is positive to Christianity and his party in Israel is unifying these two groups in one party. Why not test the New Testament as well as the Old Testament? The Jews know though their methods. According to Avi and this is his video's 2 persons were stolen, to write the Koran, on the sides (hidden) they bring various messages. Afterward, these two persons were killed. They were promised to be released (this can be seen in his books and video's).

The original Koran is quite different than the one used today, in a way that some who have noticed quit their belief in the Koran. This is not easy, and thinking perhaps that if a person is born in such land, they would perhaps also take their religion's views.

Brutal ways of stoning, and all kind of rules that from Bible point of view OT. We may not judge the people, but it is of great importance warning them and the World. This was though done in Old Testament, but New Testament is a new contract with God according to the Christian religion.

From the GP explanation these go to the Pit. No religion's view is 100% correct, he even said that he knew that some Islam religious view's, is just political way against the world power. According to his saying many knows, and they see the terrifying result of these religions. He moved to Iceland away from his homeland. It is a duty to inform the world about this result. His feedback was that Icelanders are special, and the money people elite have thought mostly about themselves.

In the peaceful time all basic of needs for all will be fulfilled. Food, clothes, house, education, healthcare. This is what ALL are waiting for said this person, so they know.

The elite takes the goods from various countries for their own benefit, and fair play is not in the picture. Mammon god is their God, and that is their aim to bring the World into slavery. (Avi Lipkin view)

The basic is that the tribe of Benjamin will be the light bearers in the End Time and while the World is in deep trouble because of false information and false religion the trouble will just increase. Of vital importance is to bring out the truth about these matters.

It is vital to make hypothesis and then the future tells what is correct and what not. The hypothesis is correct, but as FB shows some prophecy must be fulfilled and in this time 2002-2016/7 a lot has happened showing we are in the End Time. Soon better time will come after this trouble time. While thinking about this as 286 is symbolized with sin, and 288 then after that time 7 x 288 = 2016 correction is possible.

It was not until in the beginning of the year 2016. This Anti Chamber picture updated and signed by DD. When FB came here to have a conference in Grensás church, May the first 2016, he did not know either about this Anti-Chamber picture. Calculated only first the year 2016 seeing then that 0,257 into the 2016 year. Saying to friends that something would perhaps something big will happen on this day 3-4 April 2016. Panama papers were publishing that day. The result was that pr. Minister of Iceland resigned. According to my finding the land is judged first 3/10 biggest bank falls here also. Corruption is judged first. This has not been cleaned up yet, fully.

Hypothesis concerning 10 years shortening and have informed the cooperatives about that hypothesis that was correct also. Such finding was personal inspiration and insight understanding the Divine.

It is worth mentioning that the calculation for 2002-2016/7 thought the End time was finished then. As FB, headman of the Institute of Pyramidology, said that obviously the calculations were correct, but more prophecy must be fulfilled. As FB pointed out correctly and fitting to Anti Chamber calculation until 2051. At least 5 more points have been discovered.

The positive side is that 2002-2018 has proven to be correctly calculated, and now Science has possibilities to start global research concerning this matter until 2060.

16

4 BLOOD MOONS 2018 BIG BLUE MOON AND 2019

That can take a few years, but likely before 2050/60, Newton was mentioning. Newton was correct up to date which is of course outstanding. That is his calculation's concerning 23 September 2015.

GP is proving the existence of God, and basically which religions are not correct. This is the result of Pyramidology.

There is a pure minority who understands these matters. According to the GP, it is the narrow way, and Jesus Christ with his blood sacrifice opened a new way as can be seen in GP year 33 and way of life. When understood this should not be any disagreement about this matter and that is the aim of Pyramid specialists.

AR comment was that he corrected himself often with the GP.

Today specialists and today happenings.

Avi Lipkin and Dr Chuck Missler were friends and deep researcher's concerning the Bible and End times.

These outstanding specialists in the Bible such as Chuck Missler (died May 1st, 2018), who was very critical and discovered thinks concerning the mark of the Beast.

Today (10 May) Avi Lipkin is having a video message to Europe 77 Countries. He is also a politician in the old land of Israel that is positive concerning Christianity and is also in cooperation with Jews against Islam which is a World danger according to his sayings.

Worthy is mentioning the outstanding British Winston Churchill, he knew about this Pyramid message, and therefore he was sure that the British would win the war against Hitler. Sometimes he had to make difficult decisions good example is the attack against the French fleet. From a struggling point of view, this seems to be necessary to secure Britain, and they had the Stone of Destiny. Somewhere Churchill said that there must be a blind person not seeing that God has a plan for humanity.

In the Bible dreams and visions are acceptable, after often criticism and sometimes envy talking. Joseph, for example, had dreams, and Pharaoh had dreams of 7 fat cows and 7 hungry. Joseph could explain the dream. Then that happened also. He was one of 12 brothers and cheated by his half- brothers. Benjamin was not born and took no part in that.

String Theory: Dr Michio Kaku explanation, and perhaps dimensions are far more than most know about. Some people have more insight than others and see more. Just because that most people can't see ghosts, does not mean that there are not people out there that can. It is not scientific to reject matters without checking very carefully. Remembering a lady talking about what she saw colours over the head of persons. Some see that perhaps and other people not. It was not until a special tool was made that could take a picture of that everyone could see.

AR did mention[1], that there came up discussions concerning Well of Life, if it could have been made afterward, that is much later than the GP was built. The result is according to AR is nonsense. AR thinks The GP was designed about 4600 years ago.

Concerning the calculations of Ante-Chamber that is 27 April to 27 August 2019, this needs further explanation. PS was mentioning this matter that is under the Granite Leaf, in which he called "The Boss". Article from S. Laboy, Appendix C. PS talks somewhere about that "The Boss" is "will of God" (or the Lord). Then PS mentions that there is one-inch shortening under "The Boss" and therefore suggesting that something would happen drastically 27 April – 27 August 2018.

What happened is that Donald Trump president of America changed the intention of setting up by Temple institute was planning for 3 Temple in 2019 and Trump changed that and accepts Jerusalem as the capital of Israel.

Here is correct to go over the history, as mentioned 1993 there was an agreement between Israel and the Papacy concerning the Temple mount. Where Papacy has access with agreement from Israel authorities.

Trump the president of America was brought up by the rule of Jesuits, a part of the Catholic Church that was forbidden. Discussions have been on Christ lam (or Christianity and Islam) which is a mixture of Judaism, Christianity, and Islam. It is incredibly strange, to say the least, the idea of such mixture is of course not some suggestions to the will of God. To worship God and the adversary in 3rd Temple is a blasphemy to God. (According to PS this is the will of God, bringing forth the evil system)

As can be seen in Ante-Chamber picture one point is 2021,135 or 49-50 days into that year seems to be the next point.

Real peace will likely not happen until Benjamin tribe will take part in that, and according to the Bible, with Ephraim and Manasseh or Britain and America. Only time can tell, concerning these matters.

17

END TIME AND PROPHECIES

The Bible is about 1/4 Prophecies according to Pyramid specialists.

There are people who are specialists in prophecies. Mentioning three prophecies that will come through before 2060 (Newton) / 2051 (ET).

Third Temple will be established in Israel according to prophecy in the Bible, perhaps already made.

Peacetime will likely come then a few years later controlled by Manasseh (USA) Benjamin (IS) and Ephraim (UK). These are Benjamin who had the mother Rachel and Jacob father, Joseph was the brother of Benjamin and father of Manasseh and Ephraim. Ephraim was over House of Israel. The Kingdom of Judah had the tribe of Judah and Benjamin.

As FB is saying, later on, all tribes of Israel will be unified according to prophecy.

It will come to a surprise if not most of these prophecies will happen before the year 2047.

18

MAIN RESULT OF THE GREAT PYRAMID

GP seems to be "Bible in stone" Bible seems to be more correct than most can imagine.

Bible is the word of God and is inspired by Mathematical methods. Christianity is the most correct religion. In 3 places Luther church is in 3 places and showing that it is correcting the Catholic Church. Jews have the day of rest correct but rejected Jesus Christ himself. Looking at the descendant passage that is corrected by Luther (1521)

There seems to be 60° triangle, 3 x 1521.

False religion is according to the GP: Are not fitting into the picture.

Judgment is made here concerning the religion, not people here (GP result)

People have freedom of choice, and as shown in the GP time between now and around the year 2994 people can make their own choice concerning this matter. Then the last 15 years people will have time to make their own choice. If Jesus Christ is rejected, then people will have a second death. Without any pain, as some have added. This is the result of AR and specialists.

This has been proven with scientific methods for about 160 years by many, but this part in the puzzle is of importance, but only about 5% of the proof (within claims of Science). Because this last part The End Time calculation was missing the whole matter has been rejected. These specialists have been mentioned before here. GP is not only showing past but also future, and a good time is coming soon.

Professor PS discovered about the scored lines in the descending passage, and as his field was astronomy, then he found also out about to what the lines pointed. That was the 7 stars shining to the Ante-Chamber "The Boss" or the Pleiades who in the year 2141 pointed to the throne of God in the heavens. (Over the GP top) Then as can be seen dragon (Devil) star is shining down the passage as the power of the Devil over the Earth. This is clearly shown in the GP, and the Devil and his followers want as many as possible go down the Pit. 2994 then the devil will be loosed again AR because people will have freedom of choice. This must, of course, be proven in a scientific way and the GP is one of the tools.

The Bible is showing that we are born again, and people who die are born again close to where they were in last life.

Before that end, people have a choice that is only through Jesus Christ. If they reject, then they have a second death. This is confirmed in Jeremiah in Bible each person is of course judged by his or her works (this is understanding of specialists working with AR, and himself).

Benjamin tribe did not cheat his brother Joseph who later went to Egypt and came to great powers. Neither did Benjamin reject Jesus Christ.

It is rather obvious that, God chose Benjamin. This is also symbolized in the silver cup in the sack of Benjamin, who got also 5 times more.

What was noticed recently is that the Grand gallery symbolized a high way to holiness, due Crucifixion Jesus Christ opens also up to the Grand gallery through broken ramp stone.

In Ps.: 22 1-6 (Bible) Says Jesus Christ "I am a worm, and not a man[15]".

The worm is having Crimson red colour and giving seed for his children willingly. Within 3 days, he will be while as wool. Fitting into Jesus Christ sayings and life.

Also knowing that there was an agreement made between Israel and the "Holy See" Papacy in the year 1993. They would be able to control Jerusalem.

To refresh the memory, Newton said the Catholic Church would gain more power 23 September 2015 and agreement were made between Pope and the UN and USA Parliament.

Donald Trump who was brought up by Jesuits, but that angle of the Catholic Church was for a long time forbidden in the Catholic Church. To explain this matter further future will tell. Now Donald Trump is moving the American Embassy in Israel to Jerusalem, perhaps building the 3rd Temple on the back of the coin and now not anymore hidden he is perhaps introducing himself as new Cyrus and silver coin is showing.

The body is the Temple, and this is all symbolized in various ways.

Zara (Through Ireland where the Catholic Church is strong) is representing the tribe of Dan. According to new information's. Falsification of history?

Tribe of Dan is in Denmark (Likely Greece also) and perhaps also Ireland.

The USA has the number 13 in their flag and the White House has similarity to Rom church and also fitting into the system as the Catholic Church established the religion of Islam according to the information on the Internet.

As can be seen in the GP no one is coming into the High way except through Jesus Christ. This system of Trump is to unify Jews, Mormonism Islam, Catholic, some call this mix. Avi Lipkin calls this New World Order.

This is through the "Well of life", saving way of Jesus Christ. This was discovered by R. Menzies and later confirmed by PS and AR.

Abyss; In our English Bibles, the Greek word abyss [a busso] is transliterated as "abyss" (RSV "bottomless pit") in every instance except Romans 10:7, where it is translated "the deep[15)]".

19

Various hypotheses of specialists in the field

The Church is not above the Bible, the word of God. Luther had this view but did not always follow that himself, a good example is the 7 day of rest which is the Sabbath.

Newton was of the opinion that the text of the Bible was showing Jesus Christ as the son of God, not God himself. Hypothesis concerning God is 3 in 1 etc. can also be discussions. These discussions are not within this Ph.D. essay purpose.

Following the conclusion of various specialists.

JT's result is that the GP is built the angle 51°51'143'' and is fulfilling Squaring of a circle. JT's conclusion is that the GP is built under Divine influence.

Mr. RM (RM) came up with the idea that the 33 Inch triangle is fitting into Jesus Christ, and the passage system is fitting into Bible prophecy. This understanding of RM has been rejected. 153 years later I invented the same matter but discovered a few weeks later that RM had understood this long before me. Having goose bumps and understanding why he did not wait to inform the specialist PS (in the year 1865) about his findings, that the GP symbolized prophecy of the Bible. RM sent a letter to PS to Egypt, but he was already there doing research on the GP.

PS went to Egypt and did research and work concerning this matter for the rest of his life. His grave had a pyramid on top. PS had great insight to the subject, and strangely understood matters far more progress in the subject, than could be expected by one man though he was specialist as Astronomer, and therefore of importance finding of scores showing the 7 stars Pleiades and in blueprint book showing exactly when the GP is built. PS discovered concerning the Inch and then connecting correctly to Britain and America. 1 inch is = 1,001 British inch which is a part of 2002 and therefore 286 = sin. Hypothesis concerning 1881 for PS, and not discovered out until today that in PS article he is mentioning the year 1881 and even expecting something drastic to happen (this was discovered 22 April 2018).

Dr Flinders Petrie was an outstanding scientific person who did correct measurements better than most in the field. His calculations opened the eyes of the religious people, showing more correctness than known beforehand. The result of Flinders Petrie was correcting and coming up with a more detailed result. His findings actually proved in a better way God's word (according to AR).

Dr John Edgar and Morton Edgar were also very exact in their calculations and then they (Edgar family ideas) discovered about the Christ angle, confirmed by others. 26°18′9-7" Their comment about the passage system which they specialized in and the structure, and their explanation concerning the passage system, that sometimes the builders went and had the passage system sometimes more complex and in much more work than they could have done more easily.

HA died rather early 1918 in his work concerning this matter, he specialized in prophecy and looking at the matter from that angle. He noticed that PS wrongly had measured from the birth of Christ, not his crucifixion and resurrection.

DD structural engineer and cooperative of HA specialized in his field and the knowledge was first to find out about the displacement factor, was not error in the building, but an error of humanity against God himself. He found also about Anti Chamber and perhaps re-calculated but did not Draw any conclusion concerning his calculations.

AR & family. His father discovered about Isaiah 19.19-20 and it fit the GP. James Rutherford came also with the explanation that when the GP is correctly understood science and religion should have no disagreement.

AR's calculations and proving various matters from a mathematical point of view is showing his great ability for the work. Basic his findings and conclusion were correct, except End Time. AR used much time and discovered about the 5th socked with Edgar Morton, and then later had hypothesis concerning Iceland as being the Benjamin tribe lost, a good part of Benjamin tribe is still lost. He drew lines to Iceland which went over Britain and Faro islands AR understanding was correct. He AR also spent great time concerning tracing our forefathers to Adam.

AR is mentioning that in the year 2623 B.C. To be exact 2622.5282 Entrance or 22 June 2623. Fitting completely into historical information. A key year, and later mentioning that 2613 is starting off the Pyramid building, and then 2589 the GP is built. This is AR finding. AR is mentioning that in the year 1938 first was confirmed concerning scores concerning 7 stars Pleiades. These 7 stars. Shown in the GP 2623 B.C.

"The Pleiades, or Seven Stars, are referred to in the Book of Job (38:31)[15] in the Bible "Canst thou bind the sweet influences of Pleiades, or lose the bands of Orion? They are also spoken of in Amos 5:8, "Seek him that market the Seven Stars and Orion." Their names are Alcyone, Merope, Maia, Electra, Tayegeta, Sterope and Calaeno. Alcyone, admired from of old, is the principal stars of the cluster; it is classified Tauri by astronomers, Mazzaoth, meaning to encircle or surround, mentioned in the next verse (Job 38:32) was the Hebrew name for the Zodiac[1] ". This was a finding of PS who mentioned 2140 B.C. Later corrected to 2141, that is within +/- 2 years limits. There is no reason to refuse AR's findings, looking at other outstanding finds of his.

Sir Isaac Newton who died 1727 and famous scientist. To name his main work of Calculus groundwork for science work and finding out about light that it had all colours in it and also telescope prototype used today using a mirror. His interest in the Bible was a huge part of his life. He used more time 42 years on the Bible than on other scientific work. Among atheists, this was not science and they attacked Newton for his interest in the matter. He had many Bibles and his conclusion was that translation was some not too good. Knowing rather early in this work that Newton had tried to find out about the code behind the Bible, which he thought was certain. This is in Bible code books on the subject[30, 33].

At the beginning of the year, Fred told me (2015) "their person had calculated this before me" He did not mention who it was. Then at the beginning of that year, Newton prophesised about the Catholic Church. His date 23 September 2015 he said beforehand then the church would gain more power. This happened up to date. Newton also said beforehand that the land of Israel would be re-established in the End time, it happened in the year 1948, and also behind the Bible is mathematics. The telescope Newton invented had no extra colours on the side, as he invented usage of mirrors in the telescope. This type is also used today in huge telescopes. Newton also studied alchemy, but that is in fact for runner of Chemistry put in poetry (see the video "Newton's dark secrets").

Dr Ivan Panin did research concerning the text of the Bible finding out about various texts that it had multiplied of 7 and therefore "inspired" by God. Perhaps some text is also multiplied of 5, for example, this is another matter to do more research on. Videos have been made on the matter and can be seen on the internet as well as in books.

Reading the book "Iceland's great inheritance" AR, research started concerning this matter and soon discovered about more material both in Icelandic and foreign language.

Pálmi Einarson has been doing research concerning numbers and has it on the Internet, from a mathematical point of view. Flower of Life. In Skógar Eyjafjöllum Thordur Tómasson has been establishing a museum. Flower of life known to Icelandic people. Measurement of each person is fitting into the formula. Palmi has designed a measurement stick. (1.1618 cm) Icelanders had this knowledge, and some have the knowledge still.

The books of Charles Russell (CR) specialized in the Bible, trying to "milk out" the meaning in the great book. (CR) had agreed on the angle of PS, and got direct information from him, and published in his book.

FB was informed about Ante-Chamber picture, 2 May 2016, but he had a conference in Grensás-church Iceland 1 May 2016. Despite the rather bad quality, concerning taking up on video. This is on the internet as "Great Pyramid and the Bible 1-3" These books are so complicated that FB have been informing about this matter every now and then. Asking about the findings and Fred has confirmed in writing his view. Pyramid specialists have discovered about the churches, correct or false, not me so much. I confirm though what is fitting and what is not.

Ante-Chamber is showing End time and misunderstood by specialists.

Now in 2018 finally understanding the main picture of the GP after 26 -7 years. Having a slow mind perhaps, but this is a uniquely complex study. I remind the reader that best specialists in the world in the field, PS and Dr Flinders Petrie did not, according to AR for 20 years not understanding the cubic and had a debate concerning the matter. Looking at all these outstanding specialists, it is quite strange to me that many rejected the hypothesis of that The GP is "Bible in stone" not all but many. RM had this idea 1865 that is the passage system in the GP is fitting prophecy in the Bible. 33 inch is the years Jesus Christ lived here, so he sent the information to PS while he was in Egypt. This was understood 153 years later. Then discovered that "Well of Life" had been understood by RM first.

The Dark side has been doing an amazing job. Using his tricks and how easy it seems to trick people? His trick that he does not exist, all is fake in the Bible, etc. Not knowing how many so-called "scientists" have full-time job propaganda concerning his point of view. This is in line with the result of Pyramid specialist mentioned here. GP is a great tool correcting matters and that is the aim of Pyramid scientific views.

Religion who are wrong and shown by the GP are those who reject Jesus Christ. The passage system is showing how much Luther corrected the Catholic Church, and also in the year, 1440 printing started. This should not come as a surprise they killed most of the apostles except Johannes

(Revelation) and Christians went to the death on stage, eaten by Lions, etc.

First the promises were made through Isaac not Ismael who was the son of Hagar. Then the fruit is not kindness. Death seems to be the aim of many in this area. Not judging people of course, fanatics are supposed to be c.a. ¼ of the religion.

The Jews who rejected Jesus Christ also reject the New Testament. If they would only read and as their result concerning Kabbalah saying that behind every letter is a number. As Pythagoras said, "everything can be explained by numbers". They still reject Jesus Christ.

Jehovah Witness were in cooperation with Charles Russell, and after his death a lawyer James Rutherford, rejected the Pyramid, and now the Jehovah witness (Jehovah is one name of God). Recent information shows that Jehovah Witness, are looking more into the Bible today. GP is still a source of research. Hopefully this essay will be a stepping stone, and perhaps there are errors that is not known until much later on. Time will tell, and the main thing is that people in Science will look at the matter, and surely from different angle as mine. Different education then mine can bring different ideas and angle, but bring further information's forth.

First some of the pictures in these books could be explained in much better way. One should not take 26 years to understand the pictures. This should be done more thoroughly. The thinking is to let people have investigation on their own. Very few do such work. The angle is partly from Economical point of view.

Origin of the Universe has been in discussions for long time. Einstein, Newton, Tesla, and Georges Le Matire have shown their views for example.

In the famous picture DaVinci Code one of main actors comes with comment concerning people: "most people do not even see what is in front of their eyes." Meeting people in the Church. While asking about Revelation 12 symbol, it seems that none of priests new about that or Newton's calculations concerning 23 September 2015. If the students here at the University would be asked, I would be surprised if 1% had noticed obvious symbols in the Sky or 4 blood moons, but obvious when people look at videos concerning the subject.

It is so surprising that specialist's in the field of theologian views, have not expressed their concerns relating to Revelation 12 symbol in the Sky. Also, most priests seem to have no clue about what is happening.

I like to mention DD structural engineer, specialist in this field. He was a firm believer that God did not exist. Because this is huge work to do this study, he saw that he was wrong and admitted that. The main thing is not to judge beforehand and try to see matters from more than one angle.

According to AR work. We live again and again (Jeremiah 31:17 "and Lord speaking to those children who have died says: "And there is hope in thine end, said the Lord that thy children shall come again to their won border." There we have not only the promise that the dead are coming back to Earth, but that they are respectively coming back to their own border, to their own countryside[4].

Dr Erlendur Haraldsson has similarity in his books concerning afterlife. He is likely one of the most well-known ex-Professor from University of Iceland.

There seems to be difference in Dr Erlendur Haraldsson view and AR, concerning that according to Jeremiah 31 people will be incarnated in the same family group. This needs further study, and has perhaps already been done partly at least. Author of this essay has done no research concerning this matter, and AR used 20-30 best specialists from UK, and authorities supported AR research, because PS result was that GP had special message to UK and USA.

This also means that people until now not knowing Jesus Christ will have another potential to do so until 2994. Around 15 years the evil side is turned loose and people can make their own judgment. These calculations made by AR are simply informing here about his basic result.

This result means also that people have "freedom" to make their own choice. If they reject Jesus Christ they will simply die and not be reborn. This is explained in the AR book. People will have their second death, rejecting Jesus Christ. This method is fair, people are often brought up in certain religion and very few are thinking scientific. Some have died and have never heard the name of Jesus or about Christianity.

Accepting Jesus Christ is vital in this matter. Correct and admit here is to understand the main picture, too much information given without much explanation. The centre of the GP is 33 Inch and also angle both to Jerusalem, but mostly to Bethlehem the birthplace of Jesus Christ. If you look closely at the picture below the 33 inches there is a broken new way from the Broadway passage up to the highway, or through Jesus Christ. Then fittingly he is called the saviour of humanity.

It says simply that he is paying for that sin of humanity. It does not say what kind of sin people are having, but explained that the only sinless person is Jesus Christ himself. Many people seem not to know this and think that they have done so many bad things that they have not any possibilities to be saved. This seems to be a misunderstanding. Every person must look at life and its sin. God does the judgment, and not for humans to do so.

Concerning sin, it seems to be the worst kind of sin rejecting Jesus Christ.

The Benjamin tribe. If we refresh our memory about the 12 tribes, then Joseph and Benjamin had the same mother Rachel, who Jacob intended to marry first, and then later finds out that his wife was Lea, so he had to work another 7 years for Rachel. Benjamin did not take any part in cheating his brother, but all others did try to kill Joseph. James the Justus, brother of Jesus, established the first church in

Jerusalem. He did not believe in Jesus Christ his brother until after the resurrection. The first church, established outside of the land of Israel, is in Glastonbury Britain. Joseph of Arimathea, the uncle of Christ, was a tin seller and a rich man. Apostle Paul visited Britain also and they were given land concerning Christianity.

When Moses went over the see it was Benjamin people who trusted God and went over first. These are the light-bearers of Israel. In his book, AR explains this "Iceland light to the nations". The forefathers of All Icelanders are. Abraham, Isaac Jacob Benjamin, and then later for example Joseph of Arimathea, and then later many Kings and Queens of UK, and Norwegian countries. It seems that the family of Jesus were of Benjamin tribe. Now SB is confirming even Maria Magdalena was of Benjamin tribe as all in the closest circle around Jesus Christ. SB has traced back to Adam and Eve, so confirming AR result.

Like in other scientific fields, this is what is best known today. Let others break the triple cord.

The fruits: Most Icelanders are Christians. Same is with the people in Faro Island and some portion of the people in Canada (Normans).

In AR result then Ephraim is in the UK and Manasseh is the USA. After NWO these 3 nations will be controlling the World (according to prophecies) and Israel that is all 12 tribes will lead the World into a more peaceful area. Benjamin tribe is key in this plan of God. Icelanders are no 1 in the World as a most peaceful country.

Lately 3 rd. Temple is planned in Jerusalem, perhaps already up.

It can be seen on the Internet that the Pyramids seems to be in a line, also quite possibilities of using magnetic technical methods. This is just info about what is going on.

Here is shown, AR work and his family, and also other specialist looking at the matter from the similarity angle. AR mentioning in his books Pyramidology I-IV where he is saying that in the book Pyramidology V certain matters will have discussions on the book V.

23 September 2017 Revelation 12 symbolized (and Newton's 2015 23 September) among End time calculations are final proof of the Bible. These suggestions of AR for the Pyramidology V. book are irrelevant and have no influence in the matter. The conclusion is that what AR was going to have in the Pyramidology V. book was to add exactly in measurement, and discussing various ideas concerning the matter. In the view, AR has been proving his subject, with the corrections, and adding various ideas can either been done later, but changes not the main result.

The Bible is a very complicated book, but people have 2 ways, either the book is all correct or all false. The view is that the Bible seems to be correct, and the GP of Egypt is "Bible in stone" as suggested by Pyramid specialists and many others. Bible explanation that Adam and Eve was created.

Some sayings in the Bible are symbolized, and also of importance mentioning and this was information from Fred Bins that in the Bible there are types, that are types for later coming persons. This is making this amazing book Bible even more outstanding than any could imagine. It is of importance that translations are qualified. "Lost in translations" has happened often, and the Bible has much more in store, that even specialists and priest could imagine.

Bible seems to be proven, with "Bible in stone". Work of many, in the act last part is not of least importance.

First person to use Science concerning GP research was Henry Alexander Rhind (in the year 1855) from Scotland, he died only 30 years old. "Rhind was the first to think out and deliberately put into practice the principles that converted excavation from treasure-hunt into a science[4]".

In scriptural understanding, numbers have meaning, as even in Ph.D. Essay this should be explained just as AR does in his books. This is known in Theologian field by many. Pyramid numbers are showing matters in similarity views.

There are specific books available on this subject that I suggest for the reader to look up further.

Explanation of basic numbers is the following: 1-11 explanation in Pyramid Quarterly from FB from a Biblical point of view. Whole booklet on each number. Five is one of the basic numbers of the GP and the meaning is grace. This number is in multiplied by other numbers.

20

SPIRITUAL PERFECTION AND EXPLANATION ON NUMBERS

10 is also one of the basic numbers, and well known as perfection.

12 meaning is the spiritual government.

17 (7+10 = 17) spiritual perfection and perfection.

23 urgency

286 = 2 x 11 x 13 the meaning of this number is a sin against God. Because of its importance a detailed story has to be said.

It was Professor Charles Lagrange, Astronomer at the Royal Observatory, Brussels, Belgium who discovered this 286 number and wrote about it in his book (1892):

"Sur La Concordance qui existe entre la Loi Historique de Brück, la Cronologie de la Bible et celle de la Grande

Pyramide de Chéops avec une interpretation novelle de plan prophétique de Révélation."

This book was translated into English and got the name "The GP" 1894. Professor Charles Lagrange belongs, the honour of being the discoverer of Le Déplacement Caractéristique

(Displacement Factor or 286) in the discovery has been of far more importance than Charles Lagrange had realized[4]. The conclusion scientists made concerning this 286 number was that there was an "error" in the building, and as I mention this brought me to the track solving this difficult puzzle. It was not according to the understanding until DD discovered that this number is in various places. Not knowing how many places though. The result of DD was that because this number is in many places in the GP this is not an error, but is typical for this GP. 286 is a symbolic number of the GP, and 5 meaning grace.

DD and HA also mentioning that Stonehenge has similarity in numbers, Sir Isaac Newton also mentioned that. To get more information's concerning that read, for example, D.D. Book mentioned here. DD result was that this is sin symbolized in the number 286 and Jesus Christ paid the sin for humanity, which is used as an entrance into this essay.

Later AR came to the conclusion that this finding of DD was correct and changed his views also in line with DD finding.

FB who has been inventive in his writings and he discovered that the number 143 had symbolized in the Bible as "Son of God" and 143 is half of 286.

FB found for example, Reykjavik has symbolized as Reykjavík or the meaning is Ray of the Sun. And 2 days in the year ray comes from the GP lining as Khuti which means knife, but also light HKO.

Maigun Solmunde Faro Island said to me that naming Ante-Chamber is not a good name. Looking up and AR is mentioning that also.

From this view, all ideas are valid until disproved. Some say that everyone has ideas concerning the GP. The most complex building in the world.

288 12x12 = 144 (then multiplied by 2 = 288) well known number from scriptural point of view 144.000 chosen 12.000 from each tribe.

Therefore, the hypothesis was that correction from 286 to 288 to be harmonious with God's will. Multiplied by 7 and we have the year 2016.

This is also confirmed online Elo8 Hugi.is put in 222222 /777 = 286 in the year 2005.

Fred said that Newton hypothesis of End Time was 2000 – 2060 he had changed his numbers a few times for example 2050 changed to 2060 and also later in Wikipedia he changed also later in his life from 2000 to 2016 which is fitting into the calculation to the year.

Rather primitive calculations compared to those of Sir Isaac Newton and AR. They calculate up to date. It is not until the beginning of the year 2016 the Ante-Chamber picture and comparing to the calculations in the year 2016 and seeing 2016,257 and therefore mentioning with HKO who can confirm this because he wrote a note long before that date. On this exact date 3-4 April, Panama Papers were globally in the media as one can look up.

2 x 144 = 288 which has a meaning as note D. Pythagoras informed about that notes have frequency and therefore numbers behind the note.

1335 waiting days

1260 7 x 180 or 3.5 x 360 years.

1881 This is the length of the Grand gallery.

Among the information in the GP is the speed of light.

"The author was a U.S. Meteorologist, and as we could expect from one of his profession, the work is largely technical and chronological. His introduces inter alia, Pyramid theories regarding the velocity of light, the equation of light, the constant of the aberration of light, the sun's parallax. His chronical presentation is based on 19th-century chronology, which of course now requires adjusting[4]".

The favourite of Fred Bins, Charles Russel publishing books concerning the subject. He had done deep research concerning the Bible but also was interested in the GP. Until his death, his confirmation concerning the symbolized meaning concerning the building is of importance.

To Charles T. Russell belongs the credit of solving the mystery of the GP's concerning Queen Chamber that it represents destiny of humanity. King's Chamber showing the high way to holiness.

Concerning Stonehenge and more stone monuments similarity built by the same people built these monuments both in Africa and Europe. Newton and DD took time calculating these monuments also. This is also in the book of professor Appendix C.

(Written 15 May 2018)

A few months ago, doing research concerning Ante-Chamber. Looking at the timeframe between 2016 and 2021,135 which is the next point in the Ante-Chamber. Then also reading PS notes who died 1900. Ante-Chamber picture is marked 1921 by DD. Later updated. While looking at this matter, some calculations concerning 2019 c.a. 27 April to 27 August, where made, and PS is mentioning 1-inch shortage. Something drastic will happen from 27 April -27 August 2018 to 27 April -27 August 2019. If the 3rd Temple was not set up in 2018 than 2019 is likely year at least of preparation, and 2nd Temple was crushed by Romans, and many Jews were killed in the year 70 A.C.

This matter is under "the Boss" of the Anti-Chamber which is a relatively small area. Measurements are of course not accurate at all. That was only a picture of the Ante-Chamber aria, and "the Boss" is under the Granit leaf in which PS calls "the will of the Lord". Anyway, therefore I wanted to go to Egypt to do these measurements but have not visited Egypt yet.

Informing both HKO and also FB (and Bjarni Th. Rögnvaldsson) also SO about the findings at the beginning of the 2018 year.

What happened 14 of May was that it was a 70-year celebration of re-established Israel in the land of Israel. 3rd Temple was perhaps established 14 of May 2018. Israel was re-established in the year 1948 this day 14 May.

Because the future is difficult to tell, the Temple Institute was preparing the 3 rd. Temple and next year 2019 was in discussions.

Yesterday 14 May 2018, America opened Embassy in Jerusalem, and this matter went to the United Nations as America is agreeing on Jerusalem as the capital of Israel. What is under the plan is either usage of this America Embassy as 3rd Temple itself, or it will be built within a short time.

Measurements of this Embassy must be checked, and we will see further later.

Avi Lipkin's idea was that Trump is opening the way for the 3rd Temple and then New World Order will put their man into the Temple. Israel would get part of Saudi Arabia and Islam would be banned. (this is Avi Lipkin hypothesis)

One can look at America as "the Boss" in the World. Future will tell us about this great event. In the request for Ph.D. information about the 3rd Temple was mentioned.

Here is a letter from Fred Binns since 2005 (his views can be heard on the Internet "Great Pyramid and Revelation of the Bible")

Fred had a conference in Grensás church 1 May 2016, Fred simply had an explanation of basic info concerning the GP, also his understanding concerning the Catholic Church. We had not discussed the Benjamin tribe so much at the time. FB knew about AR result concerning Iceland, but FB had a first different view.

Sir Isaac Newton who understood from reading the Bible that Israel would be re-established in the land of Israel. It seems that only AR had the same view as me. FB was more into the Charles Russell books and publishing material.

It was not until now this year (2018) that the explanations where finally agreed by FB, or at least Fred had not disagreement formally. At some point, God rejected 11 tribes, all except Benjamin. According to Fred on prophecies, all tribes will be unified at last.

There is a little doubt about what the Bible is saying. Understanding of God's word has been a problem for many. Even more difficult for people who do not believe in God.

Fred Bins mentioning that Reykjavik, is in fact Reykjavik as a ray of the Sun. Twice in the year Khuti (Kuti is well known in Icelandic as knife) or light is shown from the GP 1 November and 11 February. Benjamin tribe is the light bearer. Sigfús Elíasson is mentioning that Reykjavik is a City chosen by God himself.

Ingólfur Árnarson let "The Gods" choose for him where was the correct place to setup his home. Put out his "Ondvegissulur" which lead to Reykjavik.

From Historical point of view in the book Landsins forbearance, Dr Hrefna Róbertsdóttir is mentioning that in her book is about Skúli Magnusson governor, started industry concerning wool, and also Sculpture fabric, mentioned by Bjarni Pálsson, the first medical doctor. Dr Hrefna is mentioning that this had an effect concerning Reykjavik built up as the main city in Iceland. It could have been Isafjordur, Eyrarbakki or even Seyðisfjördur which were also real possibilities[14].

What is interesting concerning Skúli Magnússon called "father of Reykjavik" lived in a house Aðalstræti 10 where now a museum is. The line from the GP is going through this house, and a statue of Skúli in front of the house (according to Thorarinn Thorarinsson arkitekt). Also, it is interesting that the other place where Skúli lived is now the Peace light of Yoko Ono and John Lennon the famous member of the rock group "The Beatles". Now another book published Skúli fógeti father of Reykjavik Thorunn Jarla Valdimarsdottir explaining why Reykjavik became the main city in Iceland[16].

Just for a historical point of view, the Church in Viðey is the oldest one of its kind here in Iceland. Likely the first church though was built under Esja mountain. It is not (2018) clear where the first church was built exactly. Some say Esja is perhaps coming from Isaiah one of the most known prophet in the Old Testament.

Eggert Ólafsson poet and Bjarni Pálsson (first medical doctor) wrote a book about Iceland informing Icelanders about nature and the land. Bjarni who was married to the daughter Rannveig Skúladóttir, daughter of Skuli Magnússon governor who also suggested Bjarni to setup Sulphur fabric close to Krísuvík[31].

This matter was also worked out by Sveinn Palsson medical doctor no 2 and first Icelandic natural scientist and also the first one in Nordic countries. His wife was Thorunn Bjarnadottir and the granddaughter of Skuli Magnusson and daughter of Bjarni Palsson.

Sveinn Palsson was correcting a few errors in Eggert Olafsson and Bjarni Palsson outstanding book and brought various more information's concerning Iceland´s nature and people. Iceland is between the lines in the GP and these authors wrote about the land and its nature[32].

Katrín Jakobsdóttir pr. Minister said that in National museum speech. Bjarni Pálsson and Eggert Ólafsson book showed Icelanders that scientific critical thinking is necessary, while they went first up to Snæfellsjökull. People were afraid of the unknown. This comment from the pr. Minister is of course very profound and showing that some politicians are keeping themselves well informed.

The other scored line on the East part of Iceland goes through Langanes, and also through Faro islands. Through Iona in Scotland, in interesting book author, Dr Anna Ritchie on page 51 is a picture, marking foot 1 Inch = 2.54 cm 1/12 foot or 30,48 cm. the picture is showing lines that are similar to lines up in Thrihyrningur and could be a similarity to Einar Pálsson findings of marking land with mathematical calculations. There are 11 meters between the lines and Thorir Sigurbjornsson teacher showed me a picture concerning this matter. His group discovered in the year 2010. Thrihyrningur was also hiding place when danger arrived from abroad in the old time. They could go up there on horses. Similar measurements were both in Scotland and Iceland. The foot as basic measurement, or the Inch.

Showing these speculation's and also following researcher from Italy.

As Gudmundur G. Thorarinsson, wrote an interesting book mentioning AR work, an engineer, and Thorarinn Thorarinsson, an architect, have been helping researcher Gianezza from Italy reading out of paintings concerning Knight Templar and connections to Iceland.

21

QUEEN´S CHAMBER

In the Descending passage we have 1521 as Luther symbolized and also in the first Ascending passage 1522 (or 1521 +/- 2) another Luther symbolized and the third is in the Queen Chamber or 1521 which is symbolized by Luther as correcting Catholic. Here is correct to note that 1521 is complete lining between Catholicism and Luther church. Luther protested 1518 Strangely also into the Queen chamber there is 1727 which is the death year of Sir Isaac Newton and he is part taker of correcting the matter, this is the meaning of Queen Chamber according to Explanation of the GP, it is though correct to mention that it is in March 31 Newton died in that year. Here is of importance mentioning that this info is in the GP who was built according AR 2623-2589 before Christ.

This means in words that death year of Sir Isaac Newton is in the GP about 4350 years before, even to me who found this out is amazing. Dr PS revolutionized Pyramid hypothesis far more than could be expected. Who would know that just one person the Almighty? This is the result of Pyramid specialists. (Pyramidology specialists)

Appendix B [5)] mentioning UK (Ephraim), USA (Manasseh) and IS (Iceland))

What Dr Barði Guðmundsson wrote an essay about, it seems to fit into this matter, his ideas were based on that Icelanders had partly arrived from black see: Origin of the Icelanders: Barði Guðmundssons Heruli.

22

THE WAY TO PEACE

Sir Isaac Newton suggestion was to have the Bible (original text in most accurate) as basic, churches should correct their teachings. What he did is that he was doing research concerning the Bible. The older Bible is better. The aim is the same, we must try to understand God's will and lecture. Appendix C.

Dead-Sea-Scrolls have in fact confirmed also various matters concerning the Bible.

The aim is to understand the matter, and the only way to understand this matter clearly is through his word the Bible. The Bible is basic, and various religion's institution's in Christianity are based on the Bible, but how good are the translations? Then it can be seen that best translations have about 98% correct translations. One of the reasons is that in the original language has perhaps even 6 different meanings, showing the difficulties in translating properly.

When applying for Ph.D. it is mentioned that 3rd Temple is on the way. Perhaps it is already up, the same day as the American Embassy was opened 14 May 2018. Perhaps some kind of secret 3rd Temple or leading to that to be established.

It is said in the Bible that he did pay 286-286 = 0 for our sin's with his life. It is not clarified what kind of sin people are having. Apostle Paul was a murderer. Remember that, because many have ideas that their sin is too huge for them to have the opportunity for eternal life.

Many atheists reject Jesus Christ, and if they do so when this matter has been clarified and proved then they will have the second death[4]. Explanation is double death factor. But what will happen then? People will not suffer anything but simply will not live again.

Many churches seem not even try, to correct their mistakes. The Jehovah Witness people have this angle to learn and change their ideas while finding out more about the Bible. Their former leader J. Rutherford lawyer was just one of the leaders, and views are not acceptable today. The Old Testament Jehovah was one of the names of God. It is no guarantee, that the name was correctly written. This is for specialists to come to an agreement about.

Concerning other religion's they may have some correct views, as God has perhaps been using various religion to show his views, for example, Viking religion's, Hinduism and various Mythology.

The future work is knowledge of good and evil and tries to choose between these.

If people want peace the rules of God must be respected and known by all. Sin will also be made, but each person should try to do well to others, and you reap what you have been harvesting.

Everyone is responsible for himself or herself.

What some have done is to change the Bible translation's to fit their own opinion. This is a mistake, and not a small one. The original text should be provided as well as possible.

Various information, in the end, is more to open the eyes of people, not judging people. If we know the Bible is correct, then that is a great achievement for further steps ahead to World Peace.

Most of these information's are from other source and just showing the status today.

According to Divine Plan of the Ages. We are promised that people will learn right from wrong.

Information at the end of the essay is showing various opinion's. While sending in Ph.D. request and got an answer. It is mentioned in an email to the University of Iceland concerning the possibility that the 3rd Temple has been built in Israel. Perhaps the Prophecy was fulfilled on 14 May 2018.

1521 (Luther church), seems to be a partaker in cooperation with 1727, Time calculation of Newton. Only the future will tell us. We are living in interesting times, but greatly misunderstood by many, even some specialist is completely out of tune with these facts.

End Time is the most misunderstood time in the history of Earth. At the end time, we will have difficulties, and then better time will come. The marketing system is killing planet Earth, along with corruption. Then afterward, people will try to follow the Bible and not protest against God and Jesus Christ. This is saying of the Pyramid specialist's.

23

Pyramid status today (2019) and future work

As 1953 and 1979 calculated by AR & other Pyramid specialists failed people mostly rejected these findings and results. This essay aim is to change that as new findings have been proven and supporting the scientific results of these Pyramid's specialists.

Noticing 4 books that bring different views on Pyramid understanding and research. Adding them to this matter and they have stories to tell. More books have been written concerning the subject, but these were chosen in the connection with this matter. These are not Pyramidology books but showing interesting information.

First is "The Orion Mystery" Robert Bauval & Adrian Gilbert. The authors are showing that the Pyramids are in fact fitting to the star belt of Orion and the authors were marking Pyramids as "star system on Earth". Belt of Orion is fitting into the 3 Pyramid and was not in line with the others, which is the same with the stars[29].

They are also saying that shafts are pointing to certain stars, this is no new invention, and this is mentioned by PS and more Pyramid's specialists.

The authors are explaining matters from Egypt old Mythology point of view, and nothing is wrong with that. They also contacted Dr I. E. S. Edwards who was the keeper of the Egyptian Antiquities at the British Museum 1954-1974, and therefore also knew AR and Pyramid specialists.

As AR's calculations did not fit at the moment then Dr Edwards rejects the main result of AR. Other information's were mostly from Egypt, an old writing and religion's views.

Second book chosen is "God's of the New Millennium" Alan F. Alford. This is a massive book over 400 pages, and he seems to be a follower of Darwin's hypothesis. The Bible is basic rejecting Darwin hypothesis, but according to FB certain ideas of Darwin were perhaps correct. This is not within the scope of this essay.

"Around 200.000 years ago, the hominid known as Homo erectus transformed itself into Homo Sapiens with 50 pr. Cent increase in brain size" Alan Alford has no answer in his sayings.

Further, Alan Alford is discussing Darwin hypothesis as the basis for his work. According to his sayings Pyramid from 10.500 – 8000 B.C. (This seems to be commonly accepted today 2018). Neither fits GP historians nor the monument timeline itself.

According to AR. The author then goes to myth stories of old religion's views. Nothing wrong with that and a lot of interesting information's are there available. "Man, the evolutionary misfit" Author explaining various religion's in various places for example in S- America Peru and Mexico. There is a picture on page 181, linking Ararat Mountain to GP.

While looking at structures from Peru and other countries even 2-300 tons of stones and walls carefully designed and greatly underestimated. This is known but silenced until recently.

One researcher works by Ron Wyatt in intended of this essay is to look at with more critical views later on. His research seems to confirm various matters concerning the Bible. Alan Alford also mentioning Rudolf Gantenbrink's work putting small robot UPUAUT 2 up to the shafts of the GP and drilling a hole through the stone result. Alan also knew "The Orion Mystery[12)]".

The third book is: "Keeper of Genesis" Here the authors are Robert Bauval and Graham Hancock.

First: the authors adding to the former book mentioned: "The Orion Mystery" where there comes Graham Hancock instead of Adrian Gilbert.

These authors are adding to the story Edgar Cayce studies and result basic concerning the Sphinx. Authors also using program calculating backward and are having a hypothesis concerning the timing of GP buildings. Among other sayings is that the monuments are waiting for the riddle to be solved in their silence, this is of course correct. And the complexity of these monuments should not be underestimated[10].

The fourth book is "Sign in the Sky" Here is one of the former authors of "The Orion mystery" that is publishing this book alone. Adrian Gilbert seems to go back to Pyramid researcher's findings, for example, showing JT "Squaring of a circle" and going back to basic for example showing that the GP is at the centre of the Earth, and also that the Nile quadrant is in fit with the GP.[1]

Adrian also showing Cabala teachings and is mentioning the star gate, which is not a new idea, that matter needs further researcher work. The author is going also into sayings and linking to Jerusalem.

What is interesting in this book among others is page 148 showing "Base Site III Aramean: the "shake hands" position of Hercules and Commagene king." This hypothesis says that according to astrology is now fitting into End Time. Also mentioning the geometry of the vertical physics[17] showing an interesting picture showing in mathematical views.

It is correct to add here Ancient Architect's web side about, various information's also about the GP.

Brian Forester's both in Egypt and also in Peru for example. Following his research is incredibly interesting, and to see how the forefathers have been doing their work leaving a person in silence and thoughts afterward.

There are 5 main things that are within the scope of this work. Show that The GP is within all reasonable result "Bible in stone" and therefore proving the Bible within demands of science. This is the result as well. Shown through the mathematical scientific result. AR hypothesis that Icelanders are the tribe of Benjamin. We are living in the End Time this is the result and fitting into Sir Isaac Newton's calculations also. Anti-Chamber new findings. Further: Research work, after the Ph.D. defence.

The GP aim is to clarify religion's view from the Almighty. All are welcome, and today all accepting Jesus Christ can go up to the narrow way the Pyramid is showing this clearly. Though Benjamin tribe is the light bearer in the last days, as FB says.

24

This is future prophecy

"Kingdom of Israel and Kingdom of Judah through the ages of history and prophecy until the time when the prophecy of Ezekiel is fulfilled and the "stick" of Judah is joined to the "stick" of Ephraim and they become one "stick"

(Ezekiel 37:15-19) "and I will make of them one nation in the land upon the mountains of Israel; and one king shall be king to them all and they shall be no more two nations, neither shall they be divided into two kingdoms any more at all"

Ezekiel 37:22 and Hosea 1.11). This means in common words all Israel is to be unified[15]. Prophecy no 1.

3rd Temple Prophecy no 2.

Peacetime. Prophecy no 3.

Icelanders should be humble serving the Lord in the last days, up with the Light (symbolized for the truth).

"Many can conceive of nothing beyond the range of the senses; whom the curtain lifted, what astonishment is in store for the World!" Karl von Eckartshausen."

(4/4 2019) For Science: (open discussion)

Concerning future and prophecy. 3-4 April 2016 is a key date, discovered by me (the year) and in Ante-Chamber 0,257 or 3-4 April. Panama papers were informed globally, and ex pr. Minister of Iceland had to resign.

Newton calculations were 756 + 1260 = the year 2016, and according to Newton, New Millennium starts in April 2016. Newton's calculations where not realized until the year 2019 February. His explanation was that 756 was "Donation of Pepin"[35] happening that was told about in the Bible prophecy. Newton was not alone in this research, 6-7 best Bible specialist in his day (Newton died in the year 1727), so this was "pretty much agreed" that this was fitting into Bible prophecy. Newton's calculation's nor prophesies about happening in the year 756 were known.

John Taylor Mathematician wrote that the Great Pyramid is divinely inspired in the year 1859. Most Pyramid specialists do agree concerning that matter.

For a further explanation to the science community. The year 2021,135 or 19-20 February 2021 is the next point. Then 2029, 2046 and the year 2051 and the last one is Newton's 2060.

3rd Temple will be up according to prophecy, and Lost tribes unified, and peacetime will start, after a difficult time.

This 2016 3-4 April seems to be a key, and 1001 years later we have 3017 3-4 April. Ten third Millennium starts. AR calculates years 1994-2994, it is 23 years difference of the latter year.

While reading The Book of Daniel and knowing about Newton's calculations 756 + 1260 year as 2016. But in The Book of Daniel is also 1290 years. This is 30 years difference, and if we add 30 years to 2016, we have 2046 which is in Ante-Chamber, 93 years after 1953. If we take from 2021 + 30 years = the year 2051 which in fact is the last year according to calculation.

Prophecy in The Book of Daniel book says 2300 years delay, until the Time of the End that is from that date (457 B.C) we have 1844 which is the beginning of Ante-Chamber calculations.

Bible's specialists can look at these speculations. AR is discussing 15 years while evil is turned loose, and unfit soles to eternity will be destroyed by the almighty. This is -/+ 15 years.

Using insight and adding 15 to 2016 and 1001 we have 3032-3 latter is 3033 and the latter part of that reminds of 33 years of Jesus Christ which is a key person in this all. This can even be delayed to the year 3065, or even 93 years delay as in Ante-Chamber. That is from 3017. This if for future thinking.

Concerning Theology department, Mathematics and Statistics should be taught for the students to have more insight into the Bible. Mathematical department and History, as well as Statistic and Theology, are more related than perhaps is known today in the year (2019).

25

CONCERNING UNIVERSITY OF ICELAND

Symbolized Sæmundur on the seal in front of the University of Iceland. Guðjon Samuelsson architect of University of Iceland old building, was well known knowing about Geometry. Fight between good and evil. Sæmundur gave the devil a blow with the Bible book. Ásmundur explanation of thinking behind his work was symbolized with Pyramid and mathematical thinking. His explanation of his work concerning these views where informed and the difference between his own saying and art- specialists in the field had quite a difference.

Who knows the better artist or art specialist about his own work? Having 4-5 hours with the artist who explained his works for example about religion's ship which is in front of Seltjarnarnes church. Sun rays symbolized

Figure 52: University of Iceland

with Ingólfur Arnarson two pillars of Boaz and Joachim for the Temple of God. Ásmundur also used the Icelandic sagas and put into his sculptures. His rough sayings were anyone that does not understand his usage of Geometry of the Pyramid symbols, circle, box, and triangle does not understand his work". Ásmundur did not inform anyone else24, 36).

26

CONCLUSION

This essay is based on AR and Engineer James Rutherford work.

The GP is in the center of the Earth, and PS result was the author of the GP knew about the Earth, and various other information in the year c.a.2600 B.C.

Their work and basic of AR, as his son JR, came to the work in AR late years. About 20 best specialists in the field were working on this matter, as PS came to the result that the message from the GP has special meaning for Britain and America as the Pyramid inch = 1,001 British Inch.

PS result was also that the GP is "Bible in Stone" or the message of the Bible in Stone using Mathematics. After 27 years of research, this seems to be the case.

Dr Ivan Panin (Dr Degree Mathematics from Harvard) has shown that inside Bible is Mathematics.

The GP was measured and has been questioned. PS went to Egypt in the year 1865 and his result was that the measurement is 1 year = 1 inch. In the year looking ahead and looking up in the Bible the year 1914 seems to have a certain meaning, which came through.

The Mathematician JT came to the conclusion that the GP was divinely inspired, and fitting into Squaring of a circle Mathematical law, which is only fitting under the corner built 51°51′143″. JT saw that the side is showing days in the Year 365. RM suggested that the center corner 33 Inch symbolized Jesus Christ life up to date. GP is showing the message of the Bible. This was also PS, result.

In the year 1925, Morton Edgar and AR found the 5th Socket. After 12 years research AR result was that Iceland is between the lines. Dr John Edgar and Morton Edgar found Christ corner 26°18′9-7″ going through Bethlehem the birthplace of Jesus Christ.

AR has proven these matters Scientifically, and therefore is in this essay.

To AR and Engineer James Rutherford it was great disappointment that their End Time calculation were not correct. Pyramidology was rejected by many.

Mathematical facts are there, and in the year 1992 the author of this essay started to do research concerning this matter. 1993-4 hypothesis sent to FB about the End Time calculation.

2002-2016/7 and then 10 years shortening that is 2002-2006/7 something drastic should happen then. What happened 2008 is Bank fall and 3/10 of biggest Bank fall was in Iceland. Later discovered that the Bank's fell in the year 2006.

Next hypothesis was concerning the year 2016 and in the beginning of the year Ante Chamber was discovered there was update 2016,257, but the calculation in years and more accurate in the Ante Chamber. 3/4 April 2016 Panama papers were put in public this day which is also the crucifixion date of Jesus Christ almost 2000 years ago. This was a global scandal, but Icelandic pr. Minister resigned.

Third calculation ahead in time was in February 2018 and also discovered from Ante Chamber picture. Likely something concerning 3rd Temple would happen between 27. April – 27 August 2018. In all cases both FB was informed beforehand and also to the Pyramid group here in Iceland.

Embassy of America was moved to Jerusalem 14 May 2018 and this is explained more detailed in the essay. Coin was made with a picture of Donald Trump president of America and Cyrus in the front, on the backside is a picture of 3 Temple.

In the year 2017 information's came concerning that Sir Isaac Newton made calculation's also concerning End Time. The End Time calculation were confirmed to be correct, but some prophecy was to be fulfilled. When coming to Iceland in Ante-Chamber is showing to 2046/2051. This was useful.

The author of this essay calculated 286 x 7 = 2002 and 286+2= 288 x 7 = 2016.

Sir Isaac Newton calculated 2000 – 2016 (same key year and discovered 756 + 1260) – 2050/60 Explanation of Newton's calculations. The year 756 is Donation of Pepin prophecy and "pretty much agreed" by 5-6 persons not only Newton. 360 years x 3 ½ = 1260 years,

End Time.

Erlingur Thorsteinsson: 2002-2006/7 – 2046/2051
Sir Isaac Newton: 2000-2016- 2050/60

The only error found is End Time in the books of AR and Engineer James Rutherford.

I cannot confirm that all the information in Pyramidology I-IV is correct, and trusting information concerning basic and useful for the work. Future work is to correct if later errors are found in their work.

These books have massive information's.

When it is shown that the Bible is based on Mathematics and GP is based on massive information's. Looking first at these and the Pyramidology's result was that the information's were so massive that JT result was it is divinely inspired. Thinking of (1/10) 100 or about that likelihood that all the information's were included. 3 types of years, length to the Sun and Moon, speed of light, Pi with 6 numbers accuracy, etc. Today claim is more than then, that is why this was calculated. (9 May 2019 calculation up to (1/10)100 but this is just for teaching)

This had not been calculated before, and when not even taking everything into account the result is that possibilities are less than (1/10)52 which is within demands of Science. First 8 numbers in each year were used and later that was 6 numbers more accurate and confirmed.

According to hieroglyph in Egypt.

Information about 2623 B.C. 5 various info confirming AR result.

a) Father of History is unlikely not to be objective in his sayings, therefore he would not be called "father of History" Just recent info concerning Herodotus, 2 basic info concerning Ships found confirm Herodotus explanation, and his explanations concerning rooms under the GP is also correctly explained by Herodotus. His information fit to other information; his result was that the GP was under construction in the year 2622 B.C.

b) Khufu reigned 2623 B.C. to 2600 B.C. and died 7 years after the GP was closed so it is not a tomb. Confirmed that GP was built when Khufu reigned. That fits to that in the year 820 A.C. no body was found in the GP.

c) Khufu ships were found in the year 1954 next to the GP. Supposed to be the oldest ships found. They went to C-14 research in the year 1965 in Pittsburg University of America.

The result were the ships are since 2600 B.C. confirming the year.

d) PS was Astronomer and professor at Edinburg University his result was that the GP is built when the Devil star shined down the Descendance passage that was also around 2600 B.C. confirming also same period.

e) The Pyramid measurement Inch is showing the building year 2623 B.C. This measurement is correct according to the research. Confirming the same period.

The result is that AR and Engineer James Rutherford were correct.

AR result was that Icelanders were Benjamin tribe.

SB has been doing tracing that is in this essay showing that Icelanders were from Abraham- Isaac-Jacob-Benjamin-

In the year 2018 July was found in the Bible Rom 11.1., that Benjamin tribe was not rejected. This is confirmed by FB who's the result was that this has been globally misunderstood.

Mathematics, standard methods are used, and which are used methods are proven by AR.

Statistic: Every number in the GP has a likelihood of 1/10 if correct numbers in a row are 6 then the possibilities are (1/10)6. Suggestion from professor concerning this matter to calculate this accurately then a program is useful and is work for another essay perhaps Ph.D.

History: The GP is showing an accurate timeline and can be useful for historians to correct.

Theology: New findings concerning this matter is revolutionary for understanding The Bible which Pyramidology's agree on that the GP is symbolizing in stone with Mathematics.

Calculations made in years and understanding were correct in all 3 cases, and in 2016,257 using information's from the GP and is correct up to date. It is unlikely that the calculation correctly made for the future in a Ph.D. essay before. There is though possibilities this is just luck. Now Science has time to the year 2060 to confirm or reject this matter.

Dr Neil deGrasse Tyson was explaining while going over the work of Sir Isaac Newton, that his hair was raising on his back, concerning the talent of Newton who invented Calculus, Telescope, understanding of light, and it is worth informing about Newton's result that God is behind all, as the distance from the Sun to the Earth all in perfect harmony, and the reason we have life on Earth. This happened not by chance according to Newton.

To calculate future is no easy matter, but like Dr Neil deGrasse Tyson, who is astonished concerning Newton. He died 1727, and what he did is he calculated 23 September

2015 correct up to date. This calculation is simply stunning. In the view one of Newton's greatest achievement.

What few know is that Newton spent more than 50% of his research concerning the Bible. (42 years of his life)

The Bible is correct (original) and is telling the truth, much more detailed than best specialists in the field can imagine. (Newton's result)

GP is: "Bible in stone" within reasonable doubt, later on, there will come further proofs concerning this matter. Ante-Chamber is showing End Time, and perhaps later also connecting to King Chamber. There are direct links to King Chamber and Ante-Chamber[3).

Finding out is that Ante-Chamber has 93 years difference (3 x 31 and 31 means El, one of Gods names, with King's Chamber that is 1928 in King Chamber is fitting to 2021 in Ante-Chamber and therefore 1953 in King Chamber is fitting to 2046. In both cases 25 years (which is 5 x 5) that is from 1928 to 1953 or 2021 to 2046.

The reader is reminded to always look at the matter from critical point of view.

Here the Science has possibilities to do research concerning this matter until the year 2060, from 1992 this matter has been correctly understood, and calculation is correct. Science has claims concerning to repeat experiment's here this is partly proven 2000-2060. Future will tell the result.

The main result is that God exists and is explained in the Bible, which is inspired words and confirmed by the mathematical angle that few know about. The Bible is truth coming directly from God himself who is the author, using people to write his wisdom down, not even knowing about what they are in fact writing down.

Sir Isaac Newton's conclusion is that original writing is correct, but the problem begins concerning translations to other languages.

This result is that Israel is God's people, as confirmed in the Bible, and the Lost-tribes theory is correct. All tribes of Israel are rejected by God, except Benjamin tribe, and as FB has confirmed this matter has been globally misunderstood. Benjamin tribe is strongest in Iceland as seen by a 5th pillar, originally discovered by AR and Morton Edgar in the year 1925.

In the coming years to 2060, the system we used today will be changed, and prophecy will be fulfilled. We will see. NASA website, the explanation's concerning the prophecies in the Bible which have already been fulfilled.

The main explanations were as follows:

To explain this, think of a person that is in America. One silver coin is thrown anywhere in America. You have your eyes closed, then you bow once, and you pick up the silver coin. This is the possibilities that all the prophecies in the Bible have come through until then. (This explanation was on the Internet about 9 years ago) Probabilities that prophecies of the Bible are correct are very huge, but still not 100%.

As you can see in 4 blood moons, that September 23, 2017, is fulfilling Revelation 12. This is not only talking; this was shown in the Sky. Bethlehem star in the year 2015, Jupiter and Venus. And blood moons 2018 and 2019 January. To do the calculations, the GP of Giza has been used as a basic truth. Also, the Bible has to be taken seriously, since it has been proofed correctly so many times.

GP of Giza has been used as the "Pyramid of truth" as basic. In the view, we must trust the Bible, to be able to do the calculation that have already been proven to be correct 3 times. AR said that he often corrected himself with knowledge from the GP.

British Israel Federation knows about the basic result, and they have changed their website[34].

References

1. Rutherford, A. (1986). *Pyramidology Book I*. Printed Harpenden, 1st January, Great Britain. Institute of Pyramidology.

2. Rutherford, A. (1970) *Pyramidology Book II*. Printed Harpenden. 1st January 1970. Great Britain. Institute of Pyramidology.

3. Rutherford, A. & Rutherford, J. (1966). *Pyramidology Book III*. Printed Harpenden, Great Britain, 28 September 1966. Institute of Pyramidology.

4. Rutherford, A., Rutherford, J. (1986). *Pyramidology Book IV*. Printed Harpenden 1st January, Institute of Pyramidology.

5. Rutherford, (1934) *A Israel Britain: Or Anglo-Saxon Israel* (2nd ed.). Muskogee, Oklahoma, USA: Aristan Publishers.

6. Rutherford, A. (1937) *Iceland light to the nations*. Published by the author, 39, Everley Gardens Belmont, Stanmore, Mddx London.

7. Rutherford, A. (1948) *Visindaleg Opinberun*. Published by the author, 22222239, Beverly Gardens Belmont, St Anmore, MDDX London.

8. Aldersmith, H., Davidson, D. *The Great Pyramid Its Divine message*. Published originally 18 March 1925. England, reprint: Kessinger Publishing, LLC.

9. Iliffe, R. (2019). *The Newton Project*. Retrieved 2019, from http://www.newtonproject.ox.ac.uk

10. Bauval, R., Hancock, G. (1997). *Keeper of Genesis*. Published in Great Britain by William Heinemann Ltd, 81 Fulham Road London sw3 6rb.

11. Gilbert, A. (2000) *The Orion Mystery*. Cox & Wyman Ltd, Reading, Berkshire. Great Britain.

12. Alford, A.F. (1997). *Gods of the New Millennium*. McKay's of Chatham plc, Chatham, Kent. Great Britain.

13. Gudmundsson, J. (1946-53) *Dagrenning*. Reykjavik, Capital Region, Iceland: Oddi.

14. Róbertsdóttir, H. (2001). *Landsins forbetran*. Reykjavík, Capital Region, Iceland: The University of Iceland.

15. King James. (1960). *The Holy Bible*. New York America: Counsel Publishing Company.

16. Valdimarsdóttir, T. J. (2018). *Skúli fógeti father of Reykjavik*. Reykjavík, Capital Region, Iceland: JBV publishing.

17. Gilbert, A (2000). *Signs in the Sky*. Bantam Press, Great Britain.

18. Smyth, C. P. (1890, December 21). *The Corroborative Testimony of God's Stone Witness And Prophet*. Retrieved 2019, from http://www.biblestudents.com/htdbv5/htdb0105.htm

19. Rutherford, A. (1937). *Pýramídinn mikli*. Retrieved 2017, from https://endtimemanna.org/magnusson/Data/Pyramidinn-mikli-Rutherford.pdf

20. The Living Word. (2014, July 10). Ivan Panin *Bible numerics*. Retrieved 2019, from https://www.youtube.com/watch?v=sGgZRbrgR1w

21. Yale Divinity School, 1. (2015, February 25). *The Book of Revelation: Seven Signs in Heaven*. Retrieved 2019, from https://www.youtube.com/watch?v=wiu1HFn_loM

22. Adrenaline Junky. (2013). Fox News: *The Coming Four Blood Moon Tetrad Prophecy*. Retrieved 2019, from https://www.youtube.com/watch?v=afmiLRb7R-A

23. Bold Sky. (2018, January 22). *Super Blue Moon*. Retrieved 2019, from https://www.youtube.com/watch?v=2Vosxt8VYpk

24. Reykjavíkurborg. *Ásmundarsafn*. Retrieved 2019, from http://artmuseum.is/asmundarsafn

25. Wikipedia. (2017, November 3). *Coat of arms of Iceland*. Retrieved 2019, from https://en.wikipedia.org/wiki/Coat_of_arms_of_Iceland

26. Kragh, M. S. (2019). *Iceland: The Tribe Of Benjamin*. Retrieved 2019, from https://ensignmessage.com/articles/iceland-the-tribe-of-benjamin/

27. Kragh, M. S. Iceland: *The Tribe of Benjamin*. Retrieved 2019, from http://www.hope-of-israel.org/icelandben.html

28. The Red Thread (2010). *Benjamin, The Dutch People Of Belgium & Netherlands*. Retrieved 2019, from https://the-red-thread.net/Benjamin-by-Betmatrho.html

29. Bauval, R., Gilbert, A. (1994). *Signs in the Sky*. Cox & Wyman Ltd, Reading, Berkshire. Great Britain.

30. Niza310. (2011, April 7). *Newtons Dark Secrets*. Retrieved 2019, from https://www.youtube.com/watch?v=sdmhPfGo3fE

31. Ólafsson, E. (1978b). *Ferðabók Eggerts Ólafssonar og Bjarna Pálssonar um ferðir þeirra á Íslandi árin*

1752-1757. Reykjavík, Capital Region, Iceland: Örn og Örlygur.

32. Wikipedia. (2018, June 20). Sveinn Pálsson. Retrieved 2019, from https://en.wikipedia.org/wiki/Sveinn_ Pálsson

33. Giza Pyramid. *Sir Issac Newton's study of the Great Pyramid of Giza.* Retrieved 2019, from http://www. gizapyramid.com/newton.htm

34. The British-Israel-World Federation. (1918). Retrieved 2018, from http://www.britishisrael.co.uk/?fbclid =IwAR3oOahfRtWH9LnnObjdyWEVUpPhV mz0gdST009pVc7oy9saD7ohBZeVGC0

35. Wikipedia. (2019, April 3). *Donation of Pepin.* Retrieved 2019, from https://en.wikipedia.org/wiki/Donation_of_ Pepin

36. Wikipedia. (2018, September 7). *Snorri Sturluson.* Retrieved 2019, from https://is.wikipedia.org/wiki/ Snorri_Sturluson

37. Wikipedia, 1. (2018b, December 1). Great Pyramid of Giza. Retrieved 2019, from https://en.wikipedia.org/ wiki/Great_Pyramid_of_Giza

38. Pack, D,C. (17 June 2009). *America and Britain in Prophecy:*published by author.

39. Sturluson, S. (1220). *Fornaldarsögur Norðurlanda*.
    Retrieved from https://www.snerpa.is/net/forn/forn.htm

40. Bjarnason, S. (2019) *From Ahio bat Azrikam to Frigg*.
    Reykjanesbær. Unpublished.

41. Íslendingabók. (2003, January). *Ólof Sigurðardóttir*.
    Retrieved 2019, from https://www.islendingabok.is

Appendix A

1) Bible is confirming tracing also to Adam. New Testament page 1 tracing of Jesus Christ.

Pyramidology III from Adam (p. 702) References 3)

2) AR/James Rutherford. (p. 1437) References 4)

AR calculates that Adam was created in the year 5407 B.C. 3 October

The thinking is that Jesus Christ pays ransom for the Adam sin, and also for the sin of humanity.

3) Snorri Sturluson. References 36)

4) SB tracking also backs to Adam, also tracing back to Abraham-Isaac-Benjamin - then Icelanders are this link of Benjamin tribe of Israel.

1. **5. Benjamin**

בנימין
Benjamin

2. Benjamin in Hebrew
3. **Native name**
4. וְיִמְיָנֵב
5. **Pronunciation**
6. Biniamin
7. **Born**
8. 11 <u>Cheshvan</u>
10. Bela (son)
   Beker (son)
   Ashbel (son)
   Gera (son)
   Naaman (son)
   Ehi (son)
9. **Children**
   Rosh (son)
   Muppim (son)
   Huppim (son)
   Ard (son)

11. <u>[1]</u>
13. <u>Jacob</u> (father)
12. **Parents**
14. <u>Rachel</u> (mother)

16. <u>Dinah</u> (half sister)

17. <u>Reuben</u> (half brother)
    <u>Simeon</u> (half brother)
    <u>Levi</u> (half brother)
    <u>Judah</u> (half brother)
    <u>Issachar</u> (half brother)
    <u>Zebulun</u> (half brother)
    <u>Dan</u> (half brother)
    <u>Naphtali</u> (half brother)
    <u>Gad</u> (half brother)
    <u>Asher</u> (half brother)
    <u>Joseph</u> (brother)

15. **Rela**

Appendix B

1) Erlingur Thorsteinsson: "Pyramidology V" this will be the name of the Ph.D. essay.

Over viewing all basic work known, concerning Pyramidology and the GP.

These were the last Pyramid specialists doing research from this angle. That is AR and his son the engineer James Rutherford and FB.

The Pyramid is also growing plants, perhaps magnetic source and various other hypothesis concerning the building. This is not the angle in this book, but work of Rutherford. While AR was doing his research, he had around 20 Specialist's in their field and this teamwork is behind Pyramidology I-IV books

2) Tablets concerning 5 sockets in the GP.

Found pylon finders

- 1 1801 N-E Colonel Cartelle- Le Pére

- 2 1801 N- W Colonel Cartelle- Le Pére

- 3 1865 S- E Aiton – Inglish

- 4 1865 S-W Aiton – Inglish

- 5 1925 5 AR & Edgar Morton. Or 5th

3) Fulfilling the prophecies:

(5 steps are in the Queen Chamber)

- Babylonian Kingdom

- Medio – Persian

- Grecian

- Papal Rome

- Kingdom of God (stone Kingdom)

4) 7 Deadly sins and opposite to the sin. (these are from Catholic sayings)

- 1 Lust – Chastity

- 2 Bulimia -Moderation

- 3 Greed- Generosity

- 4 Laziness – Alertness

- 5 Anger -Patience

- 6 Envy – Charity

- 7 Pride – Modesty

5) Unify prophecies of Benjamin Ephraim and Manasseh.

Psalm 80:1 /Bible.

Hear us, Shepherd of Israel,

You who lead Joseph like a flock.

You who sit enthroned between the cherubim,

Shine forth 2 before Ephraim, Benjamin, and Manasseh.

Awaken you might;

Come and save us.

6) 3rd Temple setup.

7) Unify Lost Tribes of Israel.

8) Peacetime: various articles: Reference 15-7)

Appendix C

1) Samuel Laboy Ministry of Antiquities

Follow · January 16, 2016, · Edited ·

Laboy info"

"THE "STONE MESSAGE" DISCOVERED IN THE ANTECHAMBER OF THE GP

The "stone message" has been there for five millenniums, nobody has recognized it, even scholars. No satisfactory answers to its meaning, intention, or purpose have been suggested. Some investigators refer to it as a decoration, others see it as a support for ropes to handle the stone blocks; many consider them of no value or importance.

An engineer from Puerto Rico used a special geometrical design pattern of his invention, but that he says was known and used by the Egyptian engineers for design purposes, to discover the location and meaning of this stone message. His design pattern, which he called the perfect symbol, permitted him to design the entire Antechamber room in the GP. This special design tool consists of the harmonic arrangement of the figure of a circle, a triangle and a square, traced with a straightedge and a compass. Previously, he used it to design the entire geometry of the GP, the design entrance to the pyramid, the King's Chamber and its five ceilings, the Queen's Chamber, and other pyramids. All these designs are explained and illustrated in his last two books.

It is being explained by experts that this protrusion object, observed in the Antechamber, called a "construction boss", was carved in the upper north side of two granite slabs, placed one over the other at the antechamber's entrance. Really, it was not carved. The entire north face of the two slabs was cut one inch deep, leaving exposed in the vertical center of the upper slab, this protrusion. Its size is about 5 inches height, 8 inches wide and 1 inch deep. The shape of this object is unique, having his sides inclined toward its center. The remaining face of the granite slab was cut level, 1 inch all around, with the exception of the east and west sides, where they were encased in the wall.

Laboy indicates that this type of construction clearly shows that it was built with another purpose, not to tie ropes. He explained that its inclined sides toward the center of the object will not permit to attach any rope around it. Therefore, it cannot be what it is called, a "construction boss".

However, Laboy found, using the Perfect symbol, the entire design of the Antechamber, using only one number to calculate the complete design. He explained that setting the symbol's circle's diameter at 149 inches, (the antechamber's height), he determined the shape and all dimensions of the Antechamber. Like using a computer, the symbol set its length as 117.14 inches and the width as 65.58 inches. These dimensions are, in accordance with those measured at Antechamber's dimensions.

Besides, the circle, the triangle and the square of the symbol perfectly defined the exact dimensions of the wainscots at the east and west sides, the width of the granite slab and its height, the two granite sections, the portcullises, and the important Phi upper point location at the granite slab. This Phi point, based on the Symbol arrangement, is exactly located at 6 inches over the joint of the lower and upper slabs.

According to Laboy's design, the upper Phi point is the right place to locate a Seal or any other message from the Egyptian designer to future generations. Besides, it is evidence that they were using the same symbol's pattern. The protrusion (or Seal) was designed using the symbol and it is located as predicted, exactly at the Phi point. He indicates that its location gives the message that the GP was finished according to plans and specifications, and with its shape, that it represents an icon of the GP. He has accumulated enough irrefutable evidence to prove that this is a "stone message".

To corroborate this analysis, engineer Laboy revealed one of his most important discoveries concerning his Perfect Symbol. He indicates that the diameter of any circle can be easily divided into two unequal parts where the ratio between the two parts is equal to the Phi Function = 1.6180339…

There are two Phi points' locations in the circle's diameter that satisfy this condition. He called them point X and point X'. These two points can be obtained by geometry, or from a simple mathematical calculation. These two points, he said, were known by the Egyptian engineers and considered sacred. The fact that the protrusion was not carved, but that the entire surface of the slabs was cut 1 inch to leave this object at the vertical centerline, exactly 6 inches from the junction of the two slabs, confirms Laboy's work. As measured in the site, the "boss" is located at 6 inches above the joint of the two granite slabs, which exactly corresponds to the Phi point location.

Engineer Laboy considered these facts so important that he designed and constructed, using his Symbol, a wooden Seal model, exactly equal to the "boss" and sent it as a present to the previous Ministry of Antiquities in 2009,

including with it, the entire antechamber design, and other important pyramid's discoveries. He recommended to him that the "boss should be safeguard as a national treasure" for Egypt; since it is a "stone message" from the ancient Egyptian Engineers, finally found and decoded. If this Seal is destroyed, the message will never be delivered.

Engineer Laboy gives a call to engineers and Architects in the Government Egyptology field, to try to rescue and save the location of this important "Stone Message", from our ancestors to mankind. The entire design work for the

Antechamber is still available. The same template as used to design the GP (The Perfect Symbol), was used and the same procedure. Only one measurement was used (the diameter of the circle (149 inches height) and all measurements are generated from that number, It is proven by the design plan the exact shape, dimensions, and location where the stone message (the boss) will be located. Besides, what it is called a rise in the floor due to earthquakes appears to be stone steps, for the sealing blocks, to avoid being push inside the King's Chamber by the spoilers.

THE STONE MESSAGE - THE SO-CALLED THE CONSTRUCTION BOSS IN THE ANTECHAMBER OF THE GP" Samuel Laboy

2). Here is the FB unpublished article:

----- Forwarded Message ----

From: FB <101234.1734@compuserve.com>

To: Erlingur Þorsteinsson <isisiserlth@yahoo.com>

Sent: Tuesday, June 14, 2005, 5:22:21 PM

Subject: TWELVE

TWELVE

"The significance of twelve as the Kingdom number is very simply demonstrated. From its very beginning, commencing with the establishment of the twelve tribes of the Children of Israel as the physical presence of the kingdom of heaven on earth, this number is naturally everywhere present. A holy kingdom is no less. (Exodus 19:6 "And ye shall be unto me a kingdom of priests, and a holy nation . . ." The fact that the nation of Israel failed catastrophically to live up to the promised in no wise annuls it or sets it aside. All was to be fulfilled in the nation's Messiah, who assured his followers of the continued destiny. "And Jesus said unto them, Verily I say unto you, that ye which have followed me, in the regeneration when the Son of man shall sit in the throne of his glory, ye also shall sit upon twelve thrones, judging the twelve tribes of Israel." (Matthew 19:28)

Teaching them to pray to their King and Heavenly Father, "Thy kingdom come,

Thy will be done, in earth EVEN AS IT IS DONE IN HEAVEN". So that when we see the approaching kingdom in the form of the New Jerusalem coming down to earth from heaven we see both its twelve foundations and twelve gates in which are written the names of the twelve Apostles and the names of the twelve tribes (Rev. 21:12, 14). A great city twelve thousand furlongs square surrounded by towering walls one hundred and forty-four (12x12) cubits in height.

From the above, it is obvious that the usual analysis is hardly required but for the sake of completeness, the various instances in Scripture are noted below with the occasional note.

Genesis 14:4 Twelve years they served Chedorlaomer, and in the thirteenth year they rebelled.

This is the first mention of the cipher and has to do with kings and kingdoms. It is introduced into the Scriptural record in connection with the eventual capture of Lot and his subsequent rescue by Abraham.

Genesis 17:20. And as for Ishmael, I have heard thee: Behold, I have blessed him, and will make him fruitful, and will multiply him exceedingly; twelve princes shall he beget, and I will make him a great nation.

Again with "great nation" we have the idea of the kingdom and in

Genesis 25:16 we have "These are the sons of Ishmael, and these are their names, by their towns, and by their castles; twelve princes according to their nations" (kingdoms). But there is a more pertinent point to the above for in type they stand for the natural seed of the Heavenly Kingdom.

Ishmael being the seed of the bondwoman (Galatians 4) and therefore must be twelvefold.

Genesis 35:22. And it came to pass, when Israel dwelt in that land, that Reuben went and lay with Bilhah his father's concubine: and Israel heard it. Now the sons of Jacob were twelve:

We might ask here. Why at just this point, are we given this obvious and seemingly unrelated piece of information that Jacob's sons were twelve in number? Then in Gen 49, we see Reuben the eldest passed over in favour of Judah as to kingdom blessings. This is the first of a number of substitutions in and out of the twelve tribes. The firstborn of all the twelve are replaced by the whole tribe of Levi, who is then said to have no inheritance in the land so that

when the kingdom is being set up this fact is taken note of as we read. Joshua *14:4 "For the children of Joseph were two tribes, Manasseh and Ephraim: therefore, they gave no part unto the Levites in the land, save cities to dwell in, with their suburbs for their cattle and for their substance." Here Manasseh and Ephraim are substituted for Levi and Joseph".*

Likewise, if we turn to Genesis 49, we seem to find a reason for the omission of Dan from the names of the tribes in Revelation 7 his name being replaced by, again one of Joseph's sons, Manasseh. (17) "Dan shall be a serpent by the way, an adder in the path, that bitten the horse heels, so that his rider shall fall backward".

What is interesting in theses and the other places where the tribes are named despite the omission's substitutions are made to maintain the number as twelve. The last example of Dan is of particular note, "a serpent by the way, an adder in the path, that bitten the horse heels, so that his rider shall fall backward", seems to foreshadow the inclusion of Judas Iscariot amongst the Twelve Disciples (John 6:70 Jesus answered them, Have not I chosen you twelve, and one of you is a devil?) and the need to select a substitute to maintain the number (Acts 1:20 For it is written in the book of Psalms, Let his habitation be desolate, and let no man dwell therein:

and his bishop prick let another take.) Whatever the case maybe it seems clear that the maintenance of this number in the natural phase of the kingdom is carefully carried forward into the spiritual phase.

With the kingdom number reasonably demonstrated as derived from the twelve tribes of Israel, it will be less necessary to expand on every occurrence. Each will be noted and just the main characteristics pointed up to where it seems of particular interest.

The last three in the book of Genesis are as follows.

Genesis 42:13; 32. The ten brothers before their brother Joseph.

Genesis 49:28 Jacob blesses his sons

Exodus 15:27. And they came to Elim, where twelve wells of water and threescore were and ten palm trees: and they encamped there by the waters.

An interesting figure for the twelve tribes and the supposed seventy judges sets up by Moses which later entered into Judaism as the Sanhedrin and mirrored the Lord's selection of, first the twelve and then the seventy Disciples?

After this first occurrence of our cipher in the Book of Exodus, the number now occurs often and naturally as the kingdom arrangements are set up amongst the twelve tribes.

It occurs a further three times in Exodus all of which are characteristic of this feature. Exodus 24:4 where Moses set up an altar.

"And Moses wrote all the words of the LORD, and rose up early in the morning, and built an altar under the hill, and twelve pillars, according to the twelve tribes of Israel", and Exodus 28:21 and 39:14 respecting the precious stones of the High priest's breastplate. "And the stones shall be with the names of the children of Israel, twelve, according to their names . . ."

Books of Numbers and Deuteronomy. The same cipher comes naturally into use.

The one text in Leviticus (24:5) respecting the shewbread "twelve cakes" obviously demonstrates this.

Ignoring the repetition of the incident of their sojourn at Elim (Numbers 33:9) and a couple of composite numbers (thirty and six thousand for example) all the texts in the book of Numbers continue to be of this character. At the conclusion of the numbering of the people, we have the statement ". . . and the princes of Israel, being twelve men: each one was for the house of his fathers (1:44). Likewise, their offerings for the tabernacle service (7:3) "And they brought their offering before the LORD, six covered wagons, and twelve oxen; a wagon for two of the princes, and for each one an ox: and they brought them before the tabernacle." So, with all their offerings (7:84, 86, 87) "This was the dedication of the altar, in the day when it was anointed, by the princes of Israel: twelve chargers of silver, twelve silver bowls, and twelve spoons of gold . . . Exodus 15:27. And they came to Elim, where twelve wells of water and threescore were and ten palm trees: and they encamped there by the waters.

An interesting figure for the twelve tribes and the supposed seventy judges sets up by Moses which later entered into Judaism as the Sanhedrin and mirrored the Lord's selection of, first the twelve and then the seventy Disciples?

After this first occurrence of our cipher in the Book of Exodus, the number now occurs often and naturally as the kingdom arrangements are set up amongst the twelve tribes.

It occurs a further three times in Exodus all of which are characteristic of this feature. Exodus 24:4 where Moses set up an altar.

"And Moses wrote all the words of the LORD, and rose up early in the morning, and built an altar under the hill, and twelve pillars, according to the twelve tribes of Israel", and Exodus 28:21 and 39:14 respecting the precious stones of the High priest's breastplate. "And the stones shall be with the names of the children of Israel, twelve, according to their names . . ."

Books of Numbers and Deuteronomy. The same cipher comes naturally into use.

The one text in Leviticus (24:5) respecting the shewbread "twelve cakes" obviously demonstrates this.

Ignoring the repetition of the incident of their sojourn at Elim (Numbers 33:9) and a couple of composite numbers (thirty and six thousand for example) all the texts in the book of Numbers continue to be of this character. At the conclusion of the numbering of the people, we have the statement ". . . and the princes of Israel, being twelve men: each one was for the house of his fathers (1:44). Likewise, their offerings for the tabernacle service (7:3) "And they brought their offering before the LORD, six covered wagons, and twelve oxen; a wagon for two of the princes, and for each one an ox: and they brought them before the tabernacle." So, with all their offerings (7:84, 86, 87) "This was the dedication of the altar, in the day when it was anointed, by the princes of Israel: twelve chargers of silver, twelve silver bowls, and twelve spoons of

The last four occurrences of twelve are in connection with the apportioning of the various inheritances in the land respecting certain cities or towns but are really not unique as different numbers of cities are allocated to various tribes or families. Nevertheless, it is still to do with the kingdom and the portion given to the tribe of Levi is of note. In particular since they had no lot in the land. Here we are told that the total number of cities allocated to the four families of Levi were, "forty and eight (4 x 12) cities with their suburbs." Joshua 21:41. Interestingly in this book devoted to the inheritance of the kingdom there are TWELVE occurrences of the kingdom number. More to the point, although not all unique in their class, all are to do with the kingdom.

There are but two Twelves in the book of the Judges which will involving the twelve tribes which would perhaps have been best understood in comparison to the cipher Eleven (coming short of the kingdom) These two occurrences are part of a most unhappy incident which takes up the better part of Chapters 19 to 21. Essentially the concubine of a Levite who was so despicably and badly treated by the Benjamite inhabitants of Gibeah that she died. As testimony and, presumably to incite justice the Levite cut up her dead body into twelve pieces and sent them into the twelve tribes. The result was that a powerful deputation was sent of all the tribes to demand that the guilty parties should be handed up.

The response of the whole tribe of Benjamin was to "gathered together from their cities to Gibeah, to go to battle against the children of Israel."(20:14) The account of the ensuing battle is long and involved, essentially at its conclusion the better part of the tribe of Benjamin was destroyed so that the men of Israel came bitterly before the Lord and said (21:3) "O LORD God of Israel, why has this come to pass in Israel, that today there should be one tribe missing in Israel?" (Hence the applicability to cipher Eleven, falling short of the kingdom). In fact, there were at least 600 men of Benjamin who had escaped "to the rock Rimmon". But it appears that most of their womenfolk had been killed in the capture of Gibeah so still the fate of the tribe seemed to hang in the balance.

The whole enterprise of restoring the tribe being now frustrated by the fact that all the Elders had, before the Lord, sworn a solemn oath never to give any of their women to the men of Benjamin because of heinous character of the first offensive committed to the hapless concubine. Yet still further complications occur in this story, for some reason not made clear in the account, the men of Jabesh Gilead now fell afoul of the assembled tribes because it was discovered that they had not supported the original mission. Now a picked force of twelve thousand is dispatched to punish the offending populace the partial effect being that "they gave them wives which they had saved alive of the women of

Jabesh Gilead ". (21:14)

The involved and far from honourable episode does not end here but this covers the inclusion of the cipher in hand. Perhaps the best commentary on this whole incident comes from the closing statement of the Book itself. Judges 21:25 In those days there was no king in Israel: Every man did that which was right in his own eyes. Israel did have a king of course, Jehovah Himself, their failure and consequent coming short of the glory of this kingdom arising from the fact that they failed to realize this.

The next three occurrences of twelve appear in 2 Samuel 2:15; 10:6; 17:1. Here again while far from unique among the many listings of armed men in numerous battles they are themselves exclusively to do with the king David and his varied fortunes as Israel's most notable monarch.

In the Book of 1 Kings all the occurrences of the cipher have kingdom connections either as implicit with its governance i.e., numbers of troops.

Or as before arising from the number of the Tribes, or simply implicit with this fact without being so stated as with the twelve oxen supporting the brazen sea or the twelve lions guarding the steps of Solomon's throne. Their references are as follows. 4:7; 4:26; 7:15; 7:25; 7:44; 10:20; 10:26; 11:30; 16:23; 18:31; 19:19;

The two occurrences in the Book of 2 Kings are simply to do with the length of a king's reign or age at accession and again while far from unique, and indeed in a minority are certainly exclusive to this application. Found in 3:1 & 21:1

The more numerous occurrences in the Book of 1 Chronicles simplifies down to just the one category. (all in chapter 25) Omitting of course two compound numbers, and a reference already covered in the portioning of the kingdom to the tribes, we have simply the new kingdom feature of the musical service of the Levities for the forthcoming arrangements respecting what would eventually be part of the temple service. More important to our subject being the fact that "the number of them, with their brethren that were instructed in the songs of the LORD, even all that were cunning, was two hundred fourscore and eight" (25:7). That is 24 x 12, this fact being made clear from the complete enumeration of every group of twelve. Verse 9 suffices to demonstrate this. "Now the first lot came forth for Asaph to Joseph: the second to Gedaliah, who with his brethren and sons were twelve"

Of the seven occurrences in the Book of 2 Chronicles the first four are a repetition of 1 Kings (1:14; 4:4; 4:15: 9:19) with the additional information in chapter 9:25 that

"Solomon had four thousand stalls for horses and chariots, and twelve thousand horsemen; whom he bestowed in the chariot cities . . .".

2 Chronicles 12:2, 3 takes us to the sadder state of kingdom affairs when "in the fifth year of king Rehoboam, Shishak king of Egypt came up against Jerusalem, because they had transgressed against the LORD, with twelve hundred chariots, and threescore thousand horsemen . . .". The final example (2 Chr. 33:1) is a repetition of 2 Kings.

At the re-establishing of the kingdom after the captivity in Babylon Ezra selected twelve priests to officiate in the rebuilt temple (Ezra 8:24) and the two other references of the cipher refer to "twelve he goats, according to the number of the tribes of Israel". Ezra 6:17; Ezra 8:35. All three examples although again arising from the fact that Israel being composed of twelve tribes are still unique and deal only with the kingdom situation.

Likewise the one reference in Nehemiah 5:14. "Moreover from the time that I was appointed to be their governor in the land of Judah, from the twentieth year even unto the two and thirtieth year of Artaxerxes the king, that is, twelve years, I and my brethren have not eaten the bread of the governor".

The next occurrence which is found in the Book of Esther does not immediately present itself as a kingdom feature although the entire book is to do with the fate of Israel. However, the subsequent result is clearly kingdom in nature Esther being made queen by king Ahasuerus, from which royal position she was able to save the whole nation and continue the course of the kingdom. The essential details are found in Esther 2:12, 13.

"Now when every maid's turn was come to go in to king Ahasuerus, after that she had been twelve months, according to the manner of the women, (for so were the days of their purification accomplished, to wit, six months with oil of myrrh, and six months with sweet odours, and with other things for the purifying of the women;) . . . . And the king loved Esther above all the women, and she obtained grace and favours in his sight more than all the virgins; so that he set the royal crown upon her head and made her queen instead of Vashti."

The one occurrence in the psalms refers to a kingdom battle the reference to which is seen in the introduction to Psalms 60:1. "To the chief Musician upon Shushaneduth, Michtam of David, to teach; when he strove with Aramnaharaim and with Aramzobah, when Joab returned, and smote of Edom in the valley of salt twelve thousand". Jeremiah 52:20, 21 is essentially a repetition of the details from 2 Kings sadly at the dismantling of the kingdom. Quite literally as regards "The two pillars, ('fillet of twelve cubits did compass') one sea, and twelve brazen bulls.

Ezekiel 43:16 reminds one of the other kingdom visions, in that case a twelve-fold foursquare city. "And the altar shall be twelve cubits long, twelve broads, square in the four squares thereof. And Ezekiel 47:13 takes us back to the first allocation of the kingdom as being the principle of the kingdom arrangement "Thus said the Lord GOD; this shall be the border, whereby ye shall inherit the land according to the twelve tribes of Israel.

The next occurrence of twelve is interesting in that it does not involve the kingdom of Israel at all, but it is certainly set within a kingdom. "At the end of twelve months he (Nebuchadnezzar) walked in the palace of the kingdom of Babylon." (Daniel 4:29)

Interestingly, though perhaps only coincidentally, our next examples, which are found in the Gospel of Matthew are also 12 in number. The application of Matthew 19:28 to the kingdom has already been made above and the fact that, as with the twelve tribes, the "coincidence" of twelve disciples with the kingdom is also unavoidable. The majority of references here and in the following gospels are of course to the disciples so we only need to inspect those that fall without this category. Verse 19:28 has been dealt with so Matthew 9:20 is next in order.

"And, behold, a woman, which was diseased with an issue of blood twelve years, came behind him, and touched the hem of his garment". Jeremiah 8:22 "Is there no balm in Gilead; is there no physician there? Why then is not the health of the daughter of my people recovered?" At last the great Physician of the kingdom had come.

Matthew 14:20 while reminding us that here again are the twelve Disciples does also point to their ministry of the kingdom message throughout the gospel age to all who hungered for the Word. "And they did all eat and were filled:

and they took up of the fragments that remained twelve baskets full."

As noted above the few occurrences of the cipher respecting battle groups are much in the minority and are only of interest as general confirmation of the kingdom symbolism. But the next reference puts this feature on a solid basis. Matthew 26:53 "Thinkest thou that I cannot now pray to my Father, and he shall presently give me more than twelve legions of angels?". Here the King Himself tells us, in terms that are entirely unique and apply directly to the kingdom, the number of His forces.

There are fifteen references in the next gospel all but one of which other than being "coincidental" to twelve disciples are a repetition of those just dealt with. The exception is Mark 5:42 "And straightway the damsel arose and walked; for she was of the age of twelve years. And they were astonished by a great astonishment. Again, in the context of the kingdom cipher, the great \Physician is seen not only bringing health but life into His kingdom.

After the references that either duplicate or "coincide" are set aside just one of the occurrences in the Gospel of Luke remain, Luke 2:42. "And when he was

twelve years old, they went up to Jerusalem after the custom of the feast." According to the custom all Jewish boys were considered to be on the threshold of manhood and so it is in this respect no surprise that the king to be is brought to our attention. But the account goes on to make it very clear that the boy Jesus thought it was appropriate to "be about my father's business", i.e. that of His Father's kingdom.

The exclusions noted above aside, again one new occurrence of the cipher is found in the six references we have in the Gospel of John. John 11:9 "Jesus answered, Are there not twelve hours in the day? If any man walks in the day, he stumbled not, because he saw the light of this world." The immediate reference is obviously to the ordering of the day, but the context is manifestly kingdom in nature. The Prince of life was simply assuring His disciples that, as the very personification of the resurrection power of the kingdom, time would not fail till He had accomplished His "father's business" on earth.

One further occurrence of a "coincident" twelve appears in the Book of Acts out of the total of five in this book. And the other four all alike have a kingdom connection. Acts 7:8 reads, for instance, "And he gave him the covenant of circumcision: and so, Abraham begat Isaac, and circumcised him the eighth day; and Isaac begat Jacob; and Jacob begat the twelve patriarchs." Stephen in this, his last witness to the gospel message, demonstrates to the Jewish Council that the gospel first preached to Abraham and founded in the twelve patriarchs was being transferred into the coming age. And interestingly the next instance of the kingdom cipher demonstrates this same feature. Acts 19:7. Early in the ministry of the Lord and at the instigation of the Baptist some of John's disciples transferred their allegiance to the advancing kingdom work, being incorporated amongst the twelve that the Lord was selecting.

Here much later in Acts (vs. 1-6), we find disciples who knew only the baptism of John having "not so much as heard of the Holy Spirit". Whereupon the apostle Paul baptize them in the name of Christ and on the laying on of his hands brings them under the power of the spirit. We are then not surprised when the account continues "And all the men were about twelve." Acts 24:1, is an interesting scripture which alluded back to a major turning point in the apostle Paul's ministry of the kingdom message ("Because that thou mayest understand, that there are yet but twelve days since I went up to Jerusalem for to worship.") From being a free agent in the Lord's ministry he became the Lord's prisoner and the author of those unique "prison epistles" which have been such a source of spiritual guidance to the church through the gospel age.

The final reference in Acts (26:7) is similar to all the earlier "coincident" examples but is worth mentioning here briefly as it shows the "gospel transfer" of the promises given to father Abraham. "Unto which promise our twelve tribes, instantly serving God day and night, hope to come. For which hope's sake, King Agrippa, was accused by the Jews.

The epistles themselves contain just two occurrences which provide an interesting introduction to the last book to of Scripture to offer this cipher, the Book of Revelation. This book was used to introduce this cipher as the kingdom number since in it was seen the kingdom's capital city, the New Jerusalem, the foundation of which were the twelve tribes and the gated of which were the twelve Apostles.

1 Corinthians 15:5 And that he was seen of Cephas, then of the twelve: James 1:1 James, a servant of God and of the Lord Jesus Christ, to the twelve tribes which are scattered abroad, greeting.

In the GP the cipher appears quite prominently and in the correct position to mark the kingdom features of the great edifice. That is in connection with the Grand Gallery running as it does from the First Ascending Passage (symbolizing the Jewish dispensation) up to the King's Chamber, (symbolizing the spiritual phase of the kingdom). And commencing from the point where the horizontal passage leads to the Queen's Chamber, (symbolizing the earthly phase of the kingdom) The length of the roof of the Grand Gallery is 1836 pyramid inches which are 153 x 12. The following from Pyramidology Book III fills out the significance.

From the level of the top of the Great Step and King's Chamber floor up to the summit platform there are 153 courses of masonry. (See also footnote on page 174 of Book I.) The significance of the wonderful number 153 greatly interests both Bible students and Pyramid students. The fact that on the memorable occasion of the great drought of fishes, recorded in John 21:11 the number is

not given in round figures as, about 150, but stated precisely as 153, infers a special reason for this and indicates that the number has a special significance. At the outset, it is interesting to note the mathematical feature that 153 is the sum of all whole numbers 1 to 17. Ie.1+2+3+4+5+6+7+8+9+ 10+11+12+13+14+15+16+17 = 153

This number, 153, is as prominent in the Pyramid as it is in Scripture.

Students of the Bible are well agreed that the 153 fish, in the text referred to above, symbolizes the true followers of Christ. Jesus said to His Apostles "Follow me and I will make you fishers of men". By the early Christians, fish were adopted as their symbol and were sometimes displayed on their tombs.

As is well known, the letters of the Greek alphabet were also used as numerals, hence Greek writing had numeric value or geometric.

In the Biblical sentence telling us of the 153 great fishes, it is remarkable that the numeric value of the Greek word for "fishes" (ichthus) is 1,224 which is precisely 8 times 153. The number 8 is the Christ number; it is the number of His name, for the gematria of Jesus in Greek (Jesus) is 888. Hence in the gematria of "fishes" in Greek, namely 1,224 (8x 153), the 8 symbolizes Christ and the 153 the chosen followers, hence it refers to the chosen followers of Christ. The precise number of days that Christ lived on Earth, from his birth on 29th September 2 B.C., till the Crucifixion on 3rd April, A.D. 33, was 12,240, which is also a multiple of 153, the factors being 10 x 8 x 153. "(FB unpublished article on number12

3. Sir Isaac Newton References 8)

The Mathematical and Scientific discoveries of Sir Isaac Newton (1642-1727) are astronomical. Some of the most notable of his achievements include the invention of calculus, the discovery of the laws of motion and the law of gravitation, and the construction of the first reflecting telescope. He also was a man known for his Christian faith. He spent a great portion of his time studying the Bible with a special interest in prophecy. Following are some of his quotations.

On the Bible:

"I have a fundamental belief in the Bible as the Word of God, written by men who were inspired. I study the Bible daily."

On atheism:

"Atheism is so senseless. When I look at the solar system. I see the earth at the right distance from the sun to receive the proper amounts of heat and light. This did not happen by chance."

Appendix D

Abraham/ Forefather of Israelites.

Adrian Gilbert Pyramid specialist.

Akbar Ezzeman document script

Al Mamoon 820 A.C. forced entrance GP

And William Orr Warden confirmed Christ angle 26°18′9.7″

Árni Björnsson historian.

Ásmundur Sveinsson sculpture artist.

Avi Lipkin author from Israel is well known for establishing a new party Jews and Christians.

Bela son of Benjamin forefathers of Icelanders.

Benjamin / One of the sons of Jacob, and therefore one of 12 tribes of Israel.

Bjarni Pálsson first medical doctor over Iceland, they wrote the first book about Iceland doing research about the land. Bjarni was married to the daughter of Skúli, Rannveig Skúladóttir, and first medical doctor museum in Seltjarnarnes.

Bobby Fischer ex-World Champion in Chess 1972 Iceland.

Captain John Maceague DD/HA

Charles Russel "Divine Plan of the Ages," wrote books concerning.

Colonel Cartelle Socket 1-2 France

Col. J. Garnier DD/HA book

Colonel Howard Vyse measuring up GP author.

DaVinci code film.

Donald Trump president of America.

Dr Alan Alford author.

Dr Anna Ritchie wrote about Iona Scotland.

Dr Bardi Gudmundsson Herule theory.

Dr Bragi Árnason professor University of Iceland and inventor concerning hydrogen.

Dr Brien Forester author of various books concerning Pyramid and also Peru.

Dr Chuck Missler specialist in the Bible

Dr Flinders Petri father of methods in Archaeology.

Dr Haraldur Nielsson professor and the rector University of Iceland.

Dr I.E.S. Edwards Keeper of Egyptian antiquities in British Museum.

Dr Ivan Panin Russian mathematical specialist from Harvard finding behind the Bible is Mathematics.

Dr John Edgar and Morton Edgar.

Dr Jon Atli Benediktsson rector of the University of Iceland.

Dr Michio Kaku a well-known scientist.

Edgar Cayce author.

Eggert Ólafsson poet and writer with Bjarni Pálsson (BP).

Einar Pálsson RIM (the root of Icelandic sagas).

Einstein inventor and scientist concerning relativity Nobel winner.

Ephraim (Britain) son of Joseph.

Fred Binns FB ex-director of institute Pyramidology UK.

George Le Maitre scientist

Gianezza Italian scientists doing research concerning Kjölur.

Graham Hancock author.

Guðjón Samúelsson architect of main building of University of Iceland.

Gudmundur G. Thorarinsson engineer and ex-president of Icelandic chess federation, and author.

Haraldur Kristján Ólason (H.K.O) cooperative policeman and inventor.

Henry Michael head man over sea map work of America discovered that GP is standing on the centre of the Quadrant of circle. Said "This monument is of the most importance in the world".

Henry Alexander Rhind first thinking to use science concerning GP.

Herodotus father historian confirmation on the building of GP 2622 B.C.

Ingólfur Arnarson first settler of Iceland (now questioned).

Isaac / was offered to God, symbolized Christ according to FB.

J. Clemenshaw (rewrote HA explaining prophecies).

James Rutherford (older) Isaiah 19.19-20

James Rutherford engineer son of AR.

Jean Francois Champollion (France) solved hieroglyphs at a young age.

Jehovah Witness religion group.

Jeremiah prophet from the Bible.

Jesus Christ born 29 Sept 2 B.C.

John Lennon peace light in Viðey was in the famous group The Beatles.

Jon Steingrímsson priest

Jónas Guðmundsson Ministry governor and author of Dagrenning.

Joseph brother of Benjamin same mother and father Jacob.

Joseph from Arimathea uncle of Jesus Christ and forefather of most Icelanders.

John Taylor JT English mathematician.

Judas apostle who cheated Jesus Christ was of Juda tribe not Benjamin.

Julian T. Gay discovered that in the GP is the speed of light.

Karl von Eckartshausen author

Katrín Jakobsdóttir pr. Minister of Iceland.

Khufu pharaoh of Egypt while GP was built.

Le Pére-Colonel Cartelle (France) Socket 1-2

L.G.A. Roberts correcting PS concerning crucifixion instead of birth.

Maigun Solmunde from Faro island and specialized in traveling.

Manasseh (America) son of Joseph.

Maria Magdalena perhaps the wife of Jesus Christ and saw him first risen from dead.

Martin Luther established the Church of Luther in the year1521.

Miller movement (year 1844).

Morton Edgar brothers, and found 5th Socket with AR

Moses from the Bible.

Nikolai Tesla inventor.

Njáll (from Njáls saga) from Icelandic sagas.

O.De.Blaere Antwerpen Belgium Isaiah 19.19-20

Pálmi Einarson designer.

Paul apostle most active apostle of them all.

Professor Charles Lagrange discovered about 286 called Displacement factor.

Professor T.E. Peet.

Piazzy Smyth PS Astronomer and professor at the University of Edinburgh.

Pythagoras Greece mathematical specialist.

Queen Elizabeth II Queen of UK since 1953.

R. Gantenbrink writer.

Rachel mother of Benjamin.

Rev commander Roberts. DD/HA

Robert Menzies RM suggested that GP was all about Jesus Christ and prophecy. To his understanding End Time was shown in Ante-Chamber, but he did not do any calculation concerning that. RM insight were outstanding.

Robert Bauval Pyramid specialist.

Ron Wyatt American scientist found various important things concerning the Bible.

Rudolf Gantenbrink Pyramid specialist put robot up to the air shafts of the GP.

Sæmundur fróði writer.

Samuel Laboy engineer and specialist concerning the GP.

Sara was Abraham wife got pregnant after such activity was "impossible".

Sigfus Eliasson poet and author.

Sigurdur Bjarnason tracing forefather's.

Sigurður Oddgeirsson teacher.

Sir Gaston Maspéro discoverer of Pyramid texts

Sir Isaac Newton famous scientist.

Sir John Maynard Keynes Economist.

Sir Winston Churchill pr. Minister of Britain during the WW2.

Skúli Magnússon governor of Iceland lived in Viðey inventor statue in Adalstraeti.

Snorri Húnbogason forefather of Icelanders.

Sr. Baldur Kristjánsson priest and bridge person.

Sr. Bjarni Rognvaldsson priest, teacher, and inventor.

Sveinn Pálsson second medical doctor over Iceland married to the daughter of Bjarni Pálsson Þórunn Bjarnadóttir and Sveinn was a first natural scientist. Statue in Vik Myrdal.

Tesla famous inventor.

Thomas Inglish engineer 3-4 Socket found

Thomas Inglish engineer also working with PS finding a 3-4 socket.

Thomas Yang (Britain) studied hieroglyphs and solved some.

Thórarinn Thórarinsson architect.

Thorir Sigurbjörnsson teacher.

Thorunn Jara Valdimarsdóttir writer of many books.

Vigdís Finnborgadóttir ex-president of Iceland.

W. Marshall Adams. DD/HA

William Aiton engineer working with PS found a 3-4 socket.

William Reeve Toronto (thought of the new horizontal scale of the GP).

Yoko Ono peace light is in Viðey. Artist.

Be critical and use science to learn. Do not judge beforehand it is unscientific.

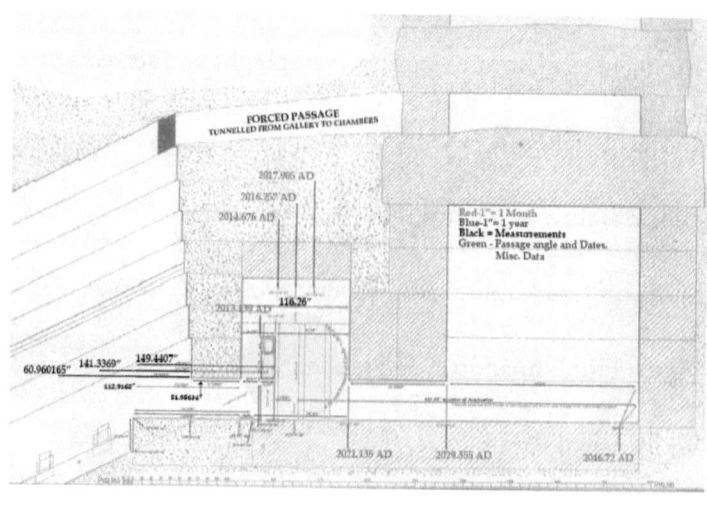

FORCED PASSAGE
TUNNELLED FROM GALLERY TO CHAMBERS

2017.905 AD

2016.357 AD

2014.676 AD

2013.197 AD   116.26"

Red-1"= 1 Month
Blue-1"= 1 year
Black = Measurements
Green - Passage angle and Dates.
Misc. Data

60.060165"   141.3369"   149.4407"

112.9168"   51.08624"

2021.139 AD   2029.555 AD   2046.72 AD